Did the Devil Make Darwin Do It?

Modern Perspectives on the Creation-Evolution Controversy

Did the Devil Make Darwin Do It?

MODERN PERSPECTIVES ON THE CREATION-EVOLUTION CONTROVERSY

EDITED BY
David B. Wilson

WITH THE ASSISTANCE OF
Warren D. Dolphin

The Iowa State University Press
AMES

To the memory of

ROBERT H. CHAPMAN

(1947–1984)

Composed and printed by
The Iowa State University Press
Ames, Iowa 50010

First edition, 1983
Second printing, 1985

Library of Congress Cataloging in Publication Data
Main entry under title:

Did the Devil make Darwin do it?

    Bibliography: p.
    Includes index.
    1. Evolution.   2. Creationism.   I. Wilson, David B.   II. Dolphin, Warren D.
QH371.D52   1983   231.7'65   82–23205
ISBN 0–8138–0433–7
ISBN 0–8138–0434–5 (pbk.)

CONTENTS

Foreword to the Second Printing, *David B. Wilson*, vii

Foreword, *Warren D. Dolphin*, viii

Introduction, *David B. Wilson*, xi

I   THE NATURE OF THE CONTROVERSY

1. Shaping Modern Perspectives: Science and Religion in the Age of Darwin,
    David B. Wilson, 3

2. A Brief Critical Analysis of Scientific Creationism,
    Warren D. Dolphin, 19

3. Theories, Facts, and Gods: Philosophical Aspects of the Creation-Evolution
    Controversy, *A. David Kline*, 37

II   THE REALM OF SCIENCE

4. Astronomical Creation: The Evolution of Stars and Planets,
    Jane L. Russell, 46

5. Interpreting Earth History, *Brian F. Glenister and Brian J. Witzke*, 55

6. The Origin of Life, *John H. Wilson*, 85

7. The Evolution of Life, *Robert H. Chapman*, 103

8. The Origin and Evolution of Humankind, *John Bower*, 114

9. Thermodynamics: The Red Herring, *Hugo F. Franzen*, 127

III   THE REALM OF RELIGION

10.   Creation Belief in the Bible and Religions, *Paul Hollenbach,* 138

11.   Christianity and Evolution, *Brent Philip Waters,* 148

12.   A Liberal Manifesto, *Henry A. Campbell,* 161

IV   CREATIONISM IN PUBLIC EDUCATION

13.   Religion in the Schools: A Historical and Legal Perspective,
       Donald E. Boles, 170

14.   The Battle in Iowa: Qualified Success,
       Jack A. Gerlovich and Stanley L. Weinberg, 189

15.   The Battle in Arkansas: The Judge's Decision, 206

Notes, 223

Index, 235

Contributors, 239

FOREWORD
TO THE SECOND PRINTING

A S this printing was being prepared in mid-1985, the heat of the creation-evolution controversy appeared to have subsided; however, the need for scholarly and scientific attention to the matter definitely had not. On the positive side, creationists had lost important battles in Arkansas, Texas, and Louisiana. In 1982, as discussed in Chapters 13 and 15, a federal district court struck down Arkansas' equal-time law. In 1984 the Texas State Board of Education's anti-evolution rule was rescinded. A harmful influence on the content of high school biology textbooks used not only in Texas but throughout the country, the decade-old rule restricted acceptable textbooks to those that presented evolution as "only one of several explanations of the origin of man" and "in a manner that is not detrimental to other theories of origin." All of the biology textbooks adopted by the Texas board in November, 1984, provided responsible discussions of evolution. In January, 1985, U.S. District Judge Adrian Duplantier struck down, without a trial, Louisiana's equal-time law. Passed by the Louisiana legislature in 1981 but never implemented, the law would have required public schools to teach the Genesis story of creation whenever evolution was taught. Judge Duplantier concluded: "Because it promotes the beliefs of some theistic sects to the detriment of others, the statute violates the fundamental First Amendment principle that a state must be neutral in its treatment of religions." In mid-1985, therefore, it no longer seemed likely that in the foreseeable future creationism would be *at the state level* an officially approved subject for high school science classes anywhere in the United States.

In Iowa, no creationism bill had been introduced in the state legislature since 1982 (see Table 14.1). However, there was concern that defeat at the legislative level had merely spurred fundamentalists to greater activity at the local level. Though it was difficult to measure precisely the influence of such activity or of opposition to it, there seemed no reason to think that the situation described in Chapter 14 (written in 1982) had greatly improved. Indeed, creationism was reportedly even being taught in high schools in two towns where colleges were located. At the same time, the Iowa Committee of Correspondence had become inactive, and the Iowa Academy of Science had disbanded its activist Panel on Controversial Issues.

In Iowa and elsewhere, the danger was that current events would repeat an earlier pattern. Though the fundamentalist view of evolution was discredited in national publicity surrounding the Scopes trial of 1925, controversy fostered by the trial helped undermine the teaching of evolution in American high schools during subsequent decades. High school teachers of the 1980s, enmeshed in other well-known problems and controversies, may understandably find it difficult to confront the creationism controversy as well. The presence of a chapter on evolution in the textbook does not automatically mean that the chapter will be taught. Whenever and wherever necessary, biology teachers deserve strong backing from their administrators and school-board members.

We welcome this printing as a chance to underscore the continued cause for concern. The absence of equal-time laws, court cases, and legislative battles is no guarantee that evolutionary theory is being adequately taught in your local high school. To find out what is being taught, ask students and teachers. And provide moral support as needed.

D. B. W.

FOREWORD

IN October 1981, while I was editing an essay for inclusion in this book, a piece of mail came to my office at Iowa State University asking if I wanted to subscribe to the newsletter of a group called Students for Origins Research, a national student society concerned with questions about creation. The flyer promised to keep me up to date on recent events, enticingly asking, "Do you know what happened when a university student was evicted from a biology classroom for questioning a professor about the theory of evolution?" How ironic this was, for in 1978 this very incident occurred in my department at Iowa State University. Actually, the student was evicted for disturbing the class, not simply for asking questions about evolution. In any case, that incident caused many of the contributors to this book to pay closer attention to the creationist movement locally and nationally.

The creationist commotion at Iowa State University extends back at least to 1976 when three creationist faculty members helped teach a special course called "Creation: A Scientific Alternative in the Study of Origins." Drawing support from conservative student Bible-study groups, the course had a large and generally enthusiastic audience. One of the faculty members told the students that though neither creation nor evolution could be proved or disproved, the truth of the creation model was established by the Bible, especially by the creation stories of Genesis and by the New Testament. Quoting from Mark 10:6, he declared that since Jesus believed in the creation stories, they must be true. Another participant in the course was a dean at Iowa State University who is on the Board of Directors of the Ann Arbor–based Creation Research Society and who is quoted in creationist literature as saying that understanding the laws of thermodynamics "is essential to the defense of biblical truths."[1]

Since 1976 Iowa and Iowa State University have witnessed the unfolding of a tragicomic play. Not only was a creationist student thrown out of class, but creationist students tried to have a biology professor thrown out of the university. Responding to remarks made by creationist students in class in previous years and to creationist literature distributed on campus, the professor discussed his critical views of creationism and fundamentalism. Upset, creationist students repeatedly distributed leaflets on the campus calling for his resignation. Their formal demand for his resignation eventually went to the Board of Regents, who rejected it. In addition, debates have been staged in the area, pitting one of the creationists' touring debaters first against an Iowa State engineering professor and next against a visiting biochemist. In five of the past six years the state legislature in nearby Des Moines has had to deal with creationist bills directed at public education. (See Table 14.1.) Faculty members have been called on or have volunteered to provide information against the proposed bills, while some students have demonstrated and testified in their favor. The local chapter of Students for Origins Research continues to be active, for example, advertising weekly classes on creationism during the 1981–82 academic year. All in all, therefore, although active creationists form only a small part of our university community, they have been vocal enough and prominent enough to make Iowa State University a major focal point in the nationwide creationist movement—or so it has seemed from our vantage point.

Initially, faculty members were reluctant to pay any serious attention to creationism. After all, other nonscientific or pseudoscientific subjects have long been with us and have not appeared to require formal renunciation by the scientific community. Astronomers and psychologists, for example, have more important things to do than to sink their time into replies to items on horoscopes and biorhythms appearing daily in newspapers. Such items seem harmless diversions taken seriously by few, it is hoped. At least, there is no movement to require high schools in the United States to teach astrology. However, when local and national events indicated that substantial numbers of people were actually taking creationism seriously and that they wanted public schools to teach it as science, some of us decided that a systematic response was necessary. After listening to creationist speakers, reading creationist writings, and talking with creationist students, we could see that creationism was nothing more than a particular version of fundamentalist Christianity having no valid scientific content. Such a conclusion could not simply be proclaimed, of course; it had to be explained. Therefore, several of the contributors to this book designed and offered a course called "Perspectives on the Creation-Evolution Controversy." The students were required to read and discuss creationist writings as well as short essays written by the instructors especially for the course. Those essays, revised and expanded, constitute much of this book.

In planning this book, the original group sought additional authors so that several perspectives on the creation-evolution controversy could be presented. The essays by educators, philosophers, ministers, a historian of science, a political scientist, and scientists from several disciplines offer viewpoints not usually found in a single volume. Our goal in writing was to explain in nontechnical terms the why, what, and how of the creation-evolution controversy as examined from several vantage points. For those beginning to examine the controversy for the first time, the popular style of the essays should allow a rapid and enjoyable venture into the realms of science, philosophy, and politics. For those who are interested in particular aspects of the controversy, each chapter contains a suggested further reading list. Those who seek a treatment of the subject that favors creationism will be disappointed. The authors are decidedly proevolution and explain why. We do not expect to convert many creationists into evolutionists. We do, however, hope to raise questions that will make creationists, evolutionists, and those uncommitted in the controversy stop and think about their positions. We hope our reasoning and opinions will help some to perceive more clearly what has become a murky subject indeed.

In closing, the encouragement of two people who have not contributed essays should be noted. Duane Isely, Distinguished Professor of Botany at Iowa State University, originally suggested that several of us publish essays derived from seminars we gave on campus. John Patterson, Professor of Materials Science and Engineering, inspired us to delve into many of these topics by his open and direct manner of questioning what is science, pseudoscience, and religion.

W. D. D.

Did the Devil Make Darwin Do It? is not intended as a facetious title. On the contrary, it sums up the sharp contrast between modern science and scholarship, on the one hand, and what is known these days as scientific creationism, on the other.

In the United States today, creationism is simply one particular version of extremely conservative Christianity. Only some conservative Christians are creationists, but evidently nearly all creationists are conservative Christians. Creationists believe that Genesis is a scientifically true account of the origins of the universe, of life on earth, and of mankind. They interpret Genesis to mean that the universe was created by God only a few thousand years ago, that the worldwide flood survived by Noah was a historical reality, and that the earth's forms of life have remained basically the same from the time of creation until now, although a few have become extinct, for example, as a result of the Noachian flood. Since the essential truths in Genesis have been revealed by God himself and reaffirmed by Jesus,[1] it follows, according to creationists, that the empirical investigations of science should be reevaluated from that religious perspective. It is in the process of such reevaluation that they propose the "scientific" portion of their viewpoint. Their usual claim is that such reevaluated science constitutes independent confirmation of the truths of the Bible.

Scientists reject creationism at various levels. They think that the Bible, as powerful a book as it is, does not provide scientific knowledge about the natural world. Nor do they think that creationists' scientific arguments stand up to the tests demanded of scientific conclusions. Many creationist arguments are totally at odds with available evidence. Underlying these areas of disagreement is the view of scientists that creationists frequently distort scientific theories, purposely misleading their audiences. Although we have not dwelt on this issue, the reader should recognize that its presence has intensified the controversy.

Creationists apparently have their own view of why evolution has gained such widespread acceptance at the expense of creationist ideas. One creationist leader has suggested that

> Satan himself is the originator of the concept of evolution. In fact, the Bible does say that he is the one "which deceiveth the whole world"

(Revelation 12:9) and that he "hath blinded the minds of them which believe not" (II Corinthians 4:4). Such statements as these must apply especially to the evolutionary cosmology, which indeed is the world-view with which the whole world has been deceived.[2]

The reader can regard this book as a three-leveled approach to the creation-evolution controversy. This introduction presents a brief overview of the controversy, while the rest of the book explores aspects of it in much greater detail. Each chapter directs the reader to an annotated reading list that could lead to comprehensive study of one or more of the scholarly and scientific disciplines represented here.

BASIC ISSUES IN THE CONTROVERSY

Some have suggested that evolutionary science (and perhaps all science) is itself religious. Like religion, the argument goes, science discusses things and events that cannot be observed. Moreover, scientists must rely on faith — faith in the conclusions reached by fellow scientists in areas of study different from their own. Let us examine the two parts of the argument separately.

First, what are science and faith? *Faith* is a word with various meanings that should be differentiated as far as possible. It can mean a set of beliefs, as in "the Christian faith." It can mean a method of knowing, as when someone says he accepts God's existence not because he can prove that God exists, but "on faith." It can mean confidence, as when we say that we have faith that the quarterback will play a good game. Thus, one might say that he has faith (confidence) that his faith (beliefs), which he accepts "on faith," will guide his daily life along the right path. With such distinctions in mind, we can see that having faith in a fellow scientist means having confidence in him or her. It does not mean that scientists in whom we have confidence accept their conclusions "on faith." Instead, scientists support their conclusions by giving reasons and by citing relevant observational and experimental evidence. Consequently, though we might have confidence in both the scientist and the theologian, if theologians accept their religious conclusions "on faith" then they accept them on quite different grounds than scientists do their conclusions, including ones about the distant past.

Second, why do scientists insist that scientific knowlege is based on observations and experiments yet so readily claim to know about the distant past, which no one has ever observed and which obviously no longer can be observed? A familiar example should help clarify the matter by illustrating the relationship between observations of the observable present and conclusions about the unobservable past. It is widely accepted, for example, that the age of a tree can be determined from the number of rings in its trunk. This correlation is based on the experience of many people with many trees in many places. Hence, if we find a tree with 400 rings, we can reasonably conclude that it has been growing for four centuries. We will not find anyone who has actually watched the tree for 400 years. We could be the first ever to see the tree.

Nonetheless, we have made a legitimate inference about the unobserved and unobservable past on the basis of present observations.

To explore the point further, imagine that someone else claims that this is a unique tree that—up to now—grew forty rings a year and is therefore only ten years old. What if he claimed, although he did not actually observe the tree's sudden appearance, that God created the tree last week with all 400 rings intact? Presumably we would reject these explanations out of hand. But why?

The essential reason is that our experience to date suggests that trees always grow in the same manner, adding one ring a year. We might, for example, confront our challenger by asking on what basis he could decide whether the tree was ten years old as he claimed, or whether it grew only ten rings a year and was thus forty years old. Or, in the case of God creating the tree, we might ask how he could tell the tree had been created a week ago and not a month ago, or how he could tell the tree had been created by one god rather than a team of gods working together, each responsible for a different component of the tree. All of these are *possible* sequences of events, and none, in principle, should be arbitrarily excluded from the realm of scientific knowledge. However, from observations of trees and reflections upon such observations, we have built up a network of understanding of trees that explains much and according to which the underlying principles of tree growth remain constant. It is possible that these principles might vary drastically, but we have no evidence that that is the case. Hence, when we fit our observation of a 400-ringed tree into this network of understanding, we reasonably conclude that it is, in fact, 400 years old.

Supernatural explanations deserve additional comment. They seem to fall into a category of *ultimate* explanations remote from the realm of science. Consider, for example, these various possibilities: nature is separate from God but God guides all natural phenomena; nature usually runs by itself, though occasionally God may help guide things; God *is* nature, and therefore natural phenomena simply represent the way God chooses to behave; nature is controlled by dozens of gods and demons, the gods responsible for events beneficial to us, the demons for those harmful. The problem is that the observable consequences of all these alternatives are the same. Whichever of the alternatives (if any) is true, natural phenomena will still appear to us as they do and therefore will provide no way for us to distinguish between or choose between competing explanations. Another view that can be said to fall into this category of ultimate explanations is materialistic determinism, according to which there are no gods, no life, no free will, and so on. Everything that happens does so in a deterministic way as a consequence only of the blind activity of matter. All our "choices" and even our sense of having a choice are actually predetermined occurrences. Our contention is that this view leads to the same observable consequences as any of the explanations involving gods. Although evolution superficially resembles materialistic determinism (neither one employs gods in its explanations), it is a mistake—one sometimes made by both creationists and evolutionists—to equate evolution and materialistic determinism. Materialistic determinism would be a far more profound explanation than evolution and, at the moment, seems as remote from science as supernatural explanations.

Perhaps we can illustrate the differences between the realms of discourse by asking the narrower question, Has life on the earth changed and, if so, how? As explained more fully in Chapter 5, the various strata of rocks in the earth's crust and their fossil contents provide the basis for answers to this question because in them is preserved a record of the changes. Most strata are sedimentary, meaning they have characteristics of sediment deposited from bodies of water. Successive layers of sediment have accumulated, mainly in the oceans, to produce bedded (layered) rocks with a cumulative thickness in any one place as great as 60,000 feet. Just as we know how long it takes a tree ring to grow, we also know about how long it takes sediment to settle from a body of water. Hence, nineteenth-century geologists concluded that it must have taken at least millions of years for these many layers to have been deposited on top of one another. In the twentieth century, radioactive dating techniques have allowed more precise dating. These techniques have extended the estimate of the earth's total age to billions of years and have confirmed the earlier judgment that the deeper strata are much older than those closer to the surface. Consequently, fossil remains contained within these strata are a record of what life was like on the earth at different times over a total period of a few billion years.

The strata and their fossil contents lead scientists to conclude that life forms on the earth have changed drastically over an immense period of time. Although details of the change from one period to another are not always clear, overall patterns of change are quite clear. Not only have many species become extinct, but species in newer strata are more advanced and more numerous than those in older strata. This does not mean there are no simple organisms still in existence, of course, but that most of today's simple organisms are different from ones that lived in the distant past and they have been joined by an increasingly greater number of increasingly advanced organisms.

Three points should be emphasized. First, observations of nature allow us to choose between the creationist's basic claim that life has existed on the earth for only a few thousand years and has stayed essentially constant during that period and the scientist's claim that life has existed on earth for an immense period and has changed greatly during that time. Indeed, though not so immediately obvious as tree rings, the "rings" of the earth and universe, once examined, yield vastly more information than those of trees. Given available evidence, deciding that the earth is only a few thousand years old, for example, is almost like saying that a tree with 400 rings is only a month old. Second, as Chapters 1 and 5 explain, long before Darwin published his *Origin of Species* in 1859, scientists already recognized that great changes had occurred on the earth. Darwin did not have to convince his fellow scientists that such changes had taken place; he was trying to convince them that he understood *how* the changes had occurred. Chapter 7 reports how the modern, updated version of Darwin's theory explains these changes and cites the evidence available to present-day scientists. Third, a scientific, evolutionary explanation of the changes in earth's life forms does not disprove the ultimate explanations mentioned earlier. For example, evolution could be guided or caused by a God or many gods, or it could be a process of materialistic determinism. Again, the point here is not whether any, or none, of these ultimate explanations is correct. Given the limitations of scientific explanation, we are currently unable to bring any of these within the domain of science. This is why Warren Dolphin says in Chapter 2 that "evolutionary theory neither implies nor denies the existence of a divine being." I would add that it also neither implies nor denies materialistic determinism.

The general concept of what does and does not constitute a scientific explanation runs through several chapters of the book and receives special attention in Chapter 3. We hope the reader can see that our distinction between the realms of science and religion is neither arbitrary nor closed to questioning. For reasons like those sketched here, this approach currently seems by far the best way to regard the problem of human knowledge.

THE CONTROVERSY VIEWED FROM SEVERAL VANTAGE POINTS

One should not imagine that the creation-evolution controversy deals only with biology and geology. Creationism runs counter to the methods and conclusions of the whole spectrum of scientific and scholarly disciplines. Of course, that is why authors from so many disciplines have contributed to this book.

Part I, "The Nature of the Controversy," contains essays by a historian of science, a biologist, and a philosopher of science. This part mainly explores general issues connected with the controversy and prepares the way for the more specific discussion in later chapters.

Chapter 1 describes how developments in religious and scientific thought in the nineteenth century's "age of Darwin" helped shape an essentially modern world view by the early part of the twentieth century. Often erroneously seen simply as an age in which progressive scientists vanquished old-fashioned clergymen, the period was actually the time of a complicated interplay between science and religion in which, for example, many scientists were Christians and many Christians adopted evolutionary views broader than Darwin's. Yet there was a marked development in thought from around 1800 to 1900. The scientists, scholars, and clergymen contributing to this book continue this nineteenth-century trend.

In Chapter 2, Warren Dolphin introduces us to modern evolutionary ideas in biology and shows how they differ from creationism. In addition, he discusses creationism as a religious and political phenomenon. Citing the main creationist journal, the *Creation Research Society Quarterly,* he documents creationists' official commitment to the literal truth of Genesis, to the status of Jesus as Lord and Savior, and to the reevaluation of science from their religious perspective. As described by Dolphin, the creationists' political attempts to introduce these views into public education in the United States gained impetus in the 1960s and, guided by various strategies, this thrust has continued to the present day.

A philosopher of science, David Kline distinguishes in Chapter 3 between the creationists' philosophical and empirical arguments. Their philosophical arguments deal with the nature of scientific knowledge generally, while their empirical arguments are concerned with the validity of specific scientific conclusions. One philosophical problem, notes Kline, is whether evolution is a fact or a theory. He argues that the term *fact* is inappropriate in the scientific context since the real issue there is how well *data* support theoretical conclusions. The crucial philosophical point is that to claim a scientific theory (like the theory of evolution) "is well established, even established beyond a reasonable doubt, is not a contradiction." Hence, the creationist's argument that evolution is only a theory does not weaken the evolutionist's position at all.

Another philosophical problem that Kline discusses involves the notion of "falsifiability," an idea proposed by the twentieth-century philosopher of science, Karl Popper, and widely adopted by scientists as a means of telling the difference between scientific and nonscientific statements. The former are said to be falsifiable, meaning they are capable of being proved wrong, or falsified, by observable evidence, or data, to use Kline's term. Popper's view points up an asymmetry in scientific knowledge regarding truth and falsehood. Although substantial negative evidence may prove a theory wrong once and for all, the same cannot be said of proving theories right. Even for theories established beyond a reasonable doubt, it remains *possible* for them to be falsified by future evidence. Although Popper's viewpoint has been criticized by other philosophers of science, it remains an important and influential insight and, as such, has become part of the creation-evolution controversy. The creationist's inconsistency on this issue is to claim both that evolution is not scientific because it is not falsifiable and also that creationists have succeeded in falsifying evolution. As Kline says, they cannot have it both ways; indeed, as this book shows, creationists have it neither way—evolution is falsifiable but has not been falsified. Kline goes on to discuss the falsifiability of creationism and the difficulty of envisioning a *super*natural god as a cause of *natural* phenomena. Having dispatched many of the philosophical issues, Kline's chapter sets the stage for the discussion of empirical matters in the scientific chapters in this book.

Part II, "The Realm of Science," contains six chapters by scientists from six different disciplines. Given Kline's analysis, what are we to make of scientists speaking of the "fact" of evolution? We need an interpretive bridge to link the philosopher's language with that of the scientist. Scientists obviously do not mean they have observed evolution for billions of years. Evolution is not, in that sense, an *observable* fact. Nor do scientists mean that evolution could not at some point in the future be falsified, since, by definition, that would make evolution nonscientific. Besides, scientists can imagine evidence that, if observed, would cause them to question evolution. The fact of evolution for scientists refers to the *occurrence* of evolution as distinguished from its *causes*. Our knowledge of the occurrence of evolution is so overwhelmingly supported by such a great amount of evidence that many scientists (perhaps most) refer to evolution as a fact in order to express their enormous confidence that it actually took place.

Kline would rather speak of evolution as a theory established beyond a reasonable doubt in order to emphasize the philosophical distinction between theory and data. We can see, therefore, that Kline and the scientists are using different terminology for the same concept so that they can call attention to the aspect in which they are most interested—the scientists in the validity of a specific conclusion, Kline in the logical structure of scientific knowledge in general. There is no basic disagreement between them. Their terminology for many of the causes of evolution (and many of the details of past evolutionary development) would no doubt be similar, for these causes are often either unknown or not known with enough certitude for scientists to speak of facts or Kline of theories established beyond a reasonable doubt. Here, Kline and the scientists would agree that we have theories that are supported in varying degrees by the evidence. The scientists, then, discuss what is currently known and not known about the evolution of the universe, and their discussion ranges from the ancient development of stars like our sun to the relatively recent emergence of humankind. They examine creationist claims in each area, show how those claims do not stand up against well-established evidence, and present an overview of modern scientific knowledge in their fields.

Chapter 4 by astronomer Jane Russell examines the arguments of a leading creationist, H. M. Morris, in his book *The Remarkable Birth of Planet Earth*. Because the universe has appeared essentially unchanged to observational astronomers during the past few thousand years, Morris rejects the conclusion of modern astronomers that the universe has changed greatly over a period of billions of years, eventually evolving into its present state. Russell explains that observational techniques available only in the last century or so reveal a universe undergoing significant change. Modern observations indicate that the universe is rapidly expanding. In retracing the course of that expansion (which can be compared to running a film backwards) astronomers have found data to support the conclusion that the expansion probably began billions of years ago in an enormous explosion. Astronomers do not claim to know what, if anything, preceded the explosion. However, they have well-substantiated ideas about how the universe developed from that long-ago compressed state to its present condition. This development is the subject of Russell's discussion, in which she explains that the evolution of stars has depended on the interaction of the force of gravity, the properties of gases, and nuclear reactions. The planets seem to be a natural by-product of the formation of the star they orbit. Hence, astronomers theorize that the earth formed at about the same time as the sun, probably a few billion years ago.

This general conclusion from astronomy fits with geological evidence, which suggests the earth is about four and a half billion years old. Geological evidence not only helps establish an age for the earth but also provides the nonscientist with what are probably the most visible and most readily understood details of the evolutionist's overall story. Conversely, it is at this point that the reader can most easily grasp the extreme shortcomings of creationists' "scientific" conclusions. Because it covers these details and their implications for creationist ideas, Chapter 5 by Brian Glenister and Brian Witzke is more extensive than the others.

Concerning the origin of life, creationists maintain that life could not have arisen naturally from nonlife, and therefore must have been created super-naturally. They argue, furthermore, that the spontaneous generation of life has never been observed and was even discredited by Pasteur during the nineteenth century. As John Wilson indicates in Chapter 6, the kind of spontaneous generation discredited by Pasteur is quite different from the processes that modern scientists think brought life into existence initially. When creationists claim that these processes are too improbable to have occurred, they are talking about processes different from those actually proposed by scientists. When crea-tionists claim that these processes are impossible because they violate the second law of thermodynamics, they fail to take into account energy changes occurring in the surroundings of developing life. Wilson goes on to explain in some detail current ideas about the origin of life. The basic task is to use modern biochemical knowledge to ascertain what chemical reactions could or would have occurred in the earth's primitive environment. Although some parts of the reconstruction are still uncertain or nonexistent, the overall picture, including many details, fits with recent experimental findings, the early fossil record, and even evidence supplied by a meteorite.

The creationists' idea of "kinds" of organisms is essential to their view of the history of earth's life. As Robert Chapman explains in Chapter 7, crea-tionists are willing to admit that change has occurred within a kind (for exam-ple, among doglike animals), but they think the kinds themselves are stable. Therefore, except for a few that have become extinct, all kinds have coexisted from the time of creation until the present day. This view is clearly refuted by the fossil record. Creationists can avoid this refutation only by maintaining that the fossil record represents, *not a chronology,* but the remains of life destroyed all at *one particular time,* the time of the Noachian Flood. Chapman re-emphasizes the point drawn in Chapter 5 that this creationist view can only be held in spite of—not because of—an overwhelming amount of evidence. The problem for modern biologists, as it was for Darwin, is to explain the changes in earth's life forms that the fossil record shows did occur. The bulk of Chap-man's chapter, then, deals with the modern theory of evolution. Although biologists disagree in some ways about how evolution took place, they agree that it did take place.

Human evolution is of special importance because of its obvious interest to humans. The general character of the discussion, however, is similar to the discussions in previous chapters. In Chapter 8, John Bower explains that in this case evidence from the fossil record is blended with current knowledge concern-ing humans and the great apes to form a coherent theory of primate evolution. The fossil record for the past twenty-five million years, though incomplete over long periods of time, still clearly illustrates the occurrence of great changes. Humans did not evolve from apes, but rather both evolved from an ancestor they had in common. In setting out the current situation in this field of research, Bower also deals with creationist objections. The evidence is certainly sufficient to overturn the creationist view that humans suddenly appeared on earth only a few thousand years ago.

Since thermodynamics forms such a fundamental part of creationist arguments, an entire chapter is devoted to the subject. Chapter 9 could well be read in conjunction with parts of others. The thermodynamics section of Chapter 1 provides an introduction to the first and second laws of thermodynamics by discussing their development in nineteenth-century physics. Chapter 2 briefly comments on the bearing of thermodynamics on organic evolution, and Chapter 6 presents a specific example to illustrate the fallacy in creationists' thermodynamic arguments concerning the origin of life. Creationists claim that evolution from disordered matter into ordered organisms and from these into organisms of ever greater order is simply impossible because it violates the second law of thermodynamics, which requires natural processes to go in the opposite direction, from order to disorder. In Chapter 9, Hugo F. Franzen points out the creationist fallacy in a discussion of the significance of the second law. He shows, for example, that according to the creationists' version of the second law, the common chemical combination of oxygen atoms into oxygen molecules could occur only rarely. If the creationists' version of the second law were correct, the oxygen in our atmosphere would exist almost entirely as oxygen atoms instead of—as is actually the case—oxygen molecules. As also pointed out by Dolphin in Chapter 2 and Wilson in Chapter 6, creationists consistently fail to consider energy changes in the environment in which chemical and biological processes occur. The result is a distorted version of the second law, which declares impossible the natural processes observed to take place around us every day. The obvious relevance of this argument to evolutionary theory is that evolutionary processes contemplated by scientists are no more thermodynamically impossible than the combination of oxygen atoms into molecules.

Part III, "The Realm of Religion," could have contained an untold number of chapters—on Islam, Judaism, agnosticism, atheism, and so on, including chapters on different factions within each of these major religions or viewpoints. However, the current creation-evolution controversy in the United States is so largely a dispute within Christianity that we have concentrated on that religion.

Actually, Chapter 10 by Paul Hollenbach is relevant to any religion involving Genesis. In a detailed analysis of the first creation story in Genesis, Hollenbach shows that a literal reading of the text yields an incomplete, inconsistent account of the world, which is contrary to much that we know about nature independently of evolution. Hence, according to Hollenbach, we should look to Genesis, not for scientific accuracy, but for religious meaning. Comparing the Genesis stories with other creation stories from other ancient cultures, he indicates something of the religious insights they provide insofar as they reflect attempts by ancient peoples to deal with universal aspects of the human situation.

In Chapter 11, Brent Waters examines a whole spectrum of Christian views of evolution during the past century. His observations indicate that not all conservative Christians have rejected evolution. Waters notes that some Christians hold the view (though not the one recommended by him) that Genesis *is* scientifically true, but that its statements must be understood in the light of the best science currently available. Thus, Genesis, when rightly interpreted, would be *evolutionary* science. This would seem to be a more profitable position for creationists to take than the one they do. It would preserve the inerrancy of Genesis, without forcing them to reject out of hand scientific accomplishments to which the rest of the world awards Nobel prizes. In any case, Waters demonstrates that the current controversy is in large part a continuation of a long-running dispute between proponents of different versions of Christianity.

Henry Campbell is a proponent of liberal Christianity, a position that in his view follows from the kind of conclusions reached in the chapters by Hollenbach and Waters. In Chapter 12, Campbell addresses the place of the liberal church in today's society. Surrounded by growing skepticism and apathy on the one side and rising fundamentalism on the other, the liberal church occupies a shrinking middle ground that its members should try to expand into a larger common ground for the common good. In Campbell's view, liberalism provides spiritual solace lacking in skepticism, without going to the extremes on intellectual and social issues often associated with narrow fundamentalism. The liberal, for example, can readily accept the modern understanding of the creation stories. Thus justified, liberalism, Campbell suggests, could greatly benefit modern American society by bringing its widely disparate elements closer together.

Over the decades the liberal Protestantism represented by Campbell has readily accepted the reality of evolution. It is therefore highly appropriate in such a broad book as this to include views like his on issues of our time. He writes from the middle of that spectrum, which goes from atheism to fundamentalism. However, the reader should not conclude that all the contributors to this volume are liberal Protestants or, as Campbell would be the first to point out, that only liberal Protestants can accept both religion and evolution. Although conservative Protestants have been more hostile to evolution than liberals, Chapter 11 shows that conservative Protestantism can accommodate evolution. One should also note the popularity of Teilhard de Chardin's views (discussed by Waters) among modern Catholics and Pope John Paul II's recent statement:

> The Bible itself speaks to us of the origin of the universe and its makeup, not in order to provide us with a scientific treatise, but in order to state the correct relationships of man with God and with the universe. Sacred scripture wishes simply to declare that the world was created by God, and in order to teach this truth it expresses itself in the terms of the cosmology in use at the time of the writer.[3]

Part IV, "Creationism in Public Education," grapples with the complex subject of public education in a pluralistic society. The complications multiply when two wholly admirable principles come into conflict, namely, that people should be able to think and believe what they want, and schools should represent the best understanding available. Chapter 13 by Donald Boles covers the history of such difficulties regarding religious teaching in schools in the United

States. Boles describes the current legal position in this area, especially as it relates to creationism. Chapter 14 by Jack Gerlovich and Stanley Weinberg focuses on the fight over creationism in Iowa in recent years. Their view—our view—is that creationists' inalienable right to believe what they want does not warrant teaching creationism as science. From that perspective, the authors recount what so far has been a successful effort in Iowa to counter creationists' legislative activities. However, state legislatures are hardly the whole of the battlefield, and Gerlovich and Weinberg indicate some of the negative effects creationism has had on the teaching of science in Iowa. They extend their chapter to the overall purpose of showing "the types of activities that the scientific, educational, legislative, and lay publics may expect from creationists as well as some mechanisms for addressing such tactics." Our concluding chapter reprints the substance of Judge William R. Overton's decision in the recent legal test of the balanced-treatment law in Arkansas. Well written and emphatic in its judgments, the decision deserves a wide readership. It strongly reinforces the content and tone of our entire book. Unfortunately, such legal defeats as the creationists met in Arkansas have not in the past stemmed their baneful influence on American science education.[4]

The Controversy and Science Education

To throw these issues into sharper relief, imagine a hypothetical example. Suppose that a group came before your local school board claiming to possess a set of multiplication tables superior to those currently used and demanding that their set be given equal time with the old set. In their tables, $5 \times 5 = 27$, $5 \times 6 = 32$, $5 \times 7 = 34$, $5 \times 8 = 41$, and so on. When asked why they accept these tables, they might give one of two answers, or a combination of both. They might say that God has revealed these tables to them and that they are therefore true. They might use strictly mathematical arguments to try to convince the school board. Or, they might say that God's revelation of the tables encouraged the pursuit of appropriate mathematical arguments and now those arguments rest on their own merits.

What would or should the school board do? Well, they might waffle, pleading there were no appropriate textbooks for this new mathematics and no teachers trained in its methods. Hence, expense and inconvenience would preclude introduction of the new system. They would, it is hoped, be more unequivocal, clearly stating that insofar as this mathematics was religion, it did not belong in a mathematics course, and that insofar as it was mathematics, it was such terrible mathematics that it certainly should be kept out of mathematics courses. Presumably, the school board would not actually decide to give the new system equal time, requiring mathematics teachers to teach both sets of multiplication tables and to say that mathematicians were unable to choose between them. Of course, everyone ought to recognize the group's right to believe that $5 \times 5 = 27$. Such tolerance should require us neither to teach students that $5 \times 5 = 27$ nor to hire a member of the group as an accountant.

Of course, the conclusions expressed in this hypothetical example apply to creationists' demand for equal time. Some may think the example unfair because 5 × 5 = 27 is so obviously incorrect that the school board's action was never in doubt. The only difference is that 5 × 5 = 27 is such elementary mathematics that we can all readily see the error, whereas we cannot *all* readily see creationists' fallacies because we are not all trained in (nor should we all be expected to be trained in) astronomy, biochemistry, geology, physics, anthropology, and so on. To those trained in these areas, creationists' "scientific" conclusions are about as obviously incorrect as 5 × 5 = 27 is to the rest of us. Thus, insofar as creationism is religion, it is one of many creation stories surviving from antiquity and as such does not belong in a science course. Insofar as it is science, it is such terrible science that it certainly should be kept out of science courses.

So starkly stated, this may simply seem an authoritarian pronouncement on behalf of the scientific community, asking for trust but giving no grounds for trust. That, of course, is why we wrote the rest of the book. As my summaries of the book's chapters try to indicate, we have sought to present a candid account of several scientific and scholarly fields bearing on the creation-evolution controversy. The chapters attempt to explain what in each field is well established, what is merely speculative, and what is totally unknown. Permeating the science chapters especially is a healthy skepticism toward the finality of scientific explanations—a conviction that all scientific conclusions are limited in scope and open to question and revision. The reader should not interpret such openness as indecisiveness. This very openness to revision demands, as Wilson so aptly states in Chapter 6, the discrimination between strong and weak explanations. To retain creationism and all other explanations ever proposed would be to refuse to revise and improve our conclusions in the light of better evidence and superior insight.

What will be the reaction to our discussion of creationism? Conceivably, we might be criticized for doing just what we say others should not—giving creationism, if not equal time, at least considerable time. I think it will be evident, however, that our discussion of creationism is not exactly what creationists have in mind when asking for equal time. In effect, we are following the suggestion that creationism *might* usefully be included in science courses—as an example of bad science. Understanding what something is *not* can often help in understanding what it is. Modern science is not creationism.

Consequently, in this book we are not merely rejecting creationism but, at a deeper level, employing it as a handy foil to help explain aspects of modern science and scholarship to a nonspecialist audience. We assembled this book for these purposes as well as in the hope that those who make, who influence, or who are interested in public policy decisions involving the creation-evolution controversy will be as well informed as possible on relevant matters.

D. B. W.

I

The Nature
of the Controversy

[1]

Shaping Modern Perspectives: Science and Religion in the Age of Darwin

DAVID B. WILSON

THE warfare between science and religion has been much exaggerated. Too often, we allow the trials of Galileo in seventeenth-century Rome and John Scopes in twentieth-century Tennessee to stand for the prevailing relationship between science and religion. In fact, the relationship has oftener been one of harmony than otherwise, and much of the conflict that has occurred has been conflict between different ways of harmonizing science and religion, rather than between science and religion themselves.

Nineteenth-century Britain—the birthplace of Charles Lyell's uniformitarian geology, Lord Kelvin's thermodynamics, and Charles Darwin's evolutionary biology—is a case in point. One should not imagine it simply as a world of clear-thinking, nonreligious scientists being opposed at every advance by fuzzy-minded, obscurantist clergymen. Lyell and Kelvin, for example, were Christians not all that enthusiastic about Darwin's ideas. Darwin, who considered himself a theist when he published his *Origin of Species* in 1859, depended upon a sophisticated biological-geological science developed mainly by Christian scientists. Many Christians adopted a version of evolution much more wide sweeping than Darwin's. Nor should one imagine that the many late Victorians who abandoned traditional Christianity did so simply, or mainly, because of evolution. For many, it was philosophical argument or biblical scholarship that rendered Christianity too implausible to deserve continued allegiance.

A much abbreviated version of this chapter forms part of an essay, "Did the Devil Make Darwin Do It? Historical Perspectives on the Creation-Evolution Controversy," *Proceedings of the Iowa Academy of Science* 89(June 1982):46–49. The author is grateful to Kenneth L. Taylor for reading this chapter and making helpful suggestions.

3

This chapter attempts to sketch such nineteenth-century developments as accurately as brevity and modern historical scholarship will allow. British thought was not the whole of nineteenth-century thought, of course, but as the immediate context of Darwin's career, it throws much light on the origin and acceptance of evolutionary ideas. Some modern evolutionists may think—or hope—that the story is simply one of scientific evolution triumphing over Christianity. Modern creationists evidently fear that is exactly what happened, so that revival of Christianity requires disproof of evolution. Such views are distortingly oversimplified. By contrast, the historian's story, sensitive to the variety of nineteenth-century thought, must as well as possible relate events as they actually happened—as they shaped an essentially modern perspective by the end of the century.

BEGINNING OF THE CENTURY

Consider the state of knowledge during the first decade or so of the nineteenth century. Detailed exploration of the fossil record had only been under way for a short time, and findings were ambiguous enough that two renowned French naturalists—Georges Cuvier and Jean Lamarck—could disagree sharply on the nature of the fossil evidence. Cuvier and others had only recently established the reality of extinction. As the preeminent European naturalist of the period, Cuvier maintained that the intricate organization of animal bodies precluded biological transmutation. Any significant deviation from the original organism would destroy the organization, resulting not in transmutation but death. Cuvier further thought that a series of localized "revolutions" or catastrophes must have been linked to the extinctions that he had recently demonstrated, for the animals involved were mobile enough to have migrated from an area in which gradual changes were occurring. British naturalists listened to Cuvier, not Lamarck whose theory of biological transmutation seemed ill supported by the evidence. Lamarck thought that an animal's internal will for improvement, coupled with the impact of the environment, produced changes in the animal during its lifetime. These "acquired characteristics" were inherited by the animal's offspring, and the accumulation of such changes led to the transmutation into a higher species.

In philosophy and theology, Britons listened to their countryman, William Paley (1743–1805), whose recently published book, *Natural Theology,* opened with a story of a man walking across a heath and coming upon a watch. Just as inspection of the watch disclosed a design that could only have been produced by an intelligent being, so also close study of the universe—especially of the earth's animal life—showed it to be the product of a designing intelligence. Paley contended that "the works of *design* are too strong to be gotten over. Design must have had a designer. That designer must have been a person. That person is GOD."[1] Thus Paley encapsulated the then familiar "Argument from design," capturing much of British thought for decades to come. In his *Evidences of Christianity,* Paley enlisted biblical miracles as the chief evidence of the truth of Christianity.

Unsurprisingly, early nineteenth-century British ideas about the earth contained a substantial theological and biblical component. The earth and its life had been designed, and the Flood remained a real event. This theological context and content were demanded by the most compelling of current philosophical-theological arguments and definitely were not forced upon unwilling scientists by church pressure. Nor did theology retard empirical studies of nature. This was the time of the foundation of the Geological Society of London, whose members were determined to organize a nationwide collection of geological data. Moreover, geological research and aspects of Enlightenment thought had by around 1800 largely undermined the earlier, more literal biblical view of an essentially unchanging earth created only six thousand years before and since then affected primarily by the Flood: "Seeing the Earth as the order imposed at the Creation and modified by the Deluge could no longer convince those familiar with the complexities uncovered by fieldwork."[2] The earth's features had been changed greatly by many agents in addition to the Flood operating in excess of a few thousand years. At the beginning of the century, therefore, there existed a unified system of theological-biblical-geological-biological thought. During the century, it was to change in many ways.

GEOLOGY: BUCKLAND AND LYELL

As geologists continued to amass enormous amounts of information concerning geological strata and their fossil contents, such information — coupled with the conclusion from physics that the earth is a continually cooling object — led them, by about 1830, to recognize progressive changes in the history of the earth and its life. The deepest, and therefore oldest, rocks seemed to have solidified from an initial molten state. Most later strata seemed to have been slowly formed by deposition of silt from bodies of water. The earliest forms of life preserved in the older rocks were fossil tropical plants, being followed by animals of increasing complexity and by plants and animals suited to moderate temperatures. The stratigraphic record indicated to most geologists that the fauna, flora, and temperature of the earth had changed considerably since its initial formation. Such matters impinged on the thought of the two leading British geologists of the 1820s and 1830s, the catastrophist William Buckland (1784–1856) and his uniformitarian student Charles Lyell (1797–1875). The careers of the two belie any simple linear development in geological-biological theory from the 1810s to the 1890s. Buckland, for example, was both more biblical and, in a way, more empirical than Lyell.

Although Buckland retained his conviction about the compatibility of Genesis and geology, he changed his views somewhat from around 1820 to the mid-1830s. Buckland thought that geological strata recorded the history of the earth *before* the events recounted at the start of Genesis. Probably representing "millions of millions of years" of earth's history, these strata told what the planet was like "in the beginning."[3] During this period, there were a number of life-destroying catastrophes followed by life-forming divine creations; the successive creations included progressively more advanced organisms. According to Buckland, the six "days" of Genesis should be interpreted as "long periods." He argued that one could not determine the geological record from the Bible, but that the geological record supported the Bible in two important ways. It confirmed the recent origin of man, and it provided empirical evidence of a recent universal flood. Buckland's strongest evidence for the latter came from cave deposits he studied in the early 1820s. By 1836, however, he had dropped this particular argument.

Lyell's uniformitarian geology appeared in his *Principles of Geology* published in three volumes in the 1830s. Hoping to remove geology even further from direct connections with the Bible than Buckland had, Lyell nevertheless placed geology in a theological context. As is probably well known, he argued that the history of the earth involved the kind of geological events we observe now (such as volcanic eruptions, earthquakes, floods, and so on), occurring in about the same degree as observed at present. He thus explained particular geological formations and showed how such small-scale events could lead to catastrophic-looking results. Moreover, and more controversially, he maintained that the earth and its life had not changed progressively but, rather, the continual change had merely oscillated around some "mean" value and there was no net change over the long run. The fossil record gave the appearance of progressive change only because it was an imperfect record.

In analyzing past and present life, Lyell emphasized the geographical distribution of plants and especially animals. The highly localized range of most animals indicated that they had been "created" in that locality, either by God or, more probably, by some natural cause as yet unknown.[4] Hence, continual geological change continually altered habitats, causing old species to die out and allowing new species to be formed. But, to repeat, the change was not progressive change. Indeed, one should expect God to have created a world perfect enough that such progressive change would be unwarranted. Moreover, the suitability of animals to their environments attested to God's design. Finally, the fossil record was complete enough to establish man's recent creation, thus setting him apart from the animal world. The Noachian Flood, however, was not part of Lyell's geology, perhaps not entirely because it would have been a nonuniformitarian event but because of the present distribution of animals. If present animals were all descendents of Noah's animals, then their populations would now be centered on or radiate out from Noah's landing place. This was decidedly not the case.

Hence, the consensus of the 1840s preserved some of Lyell's views, some of Buckland's. As Buckland and most other geologists had recognized by 1830, the earth had had a progressive history. Lyell's steady-state theory was just too contrary to the evidence to be accepted, and Lyell himself eventually abandoned it. Yet, enough of Lyell's uniformitarian ideas were accepted that the earth's progress was conceived to have occurred not during Buckland's "millions of millions of years" but during an indefinitely long period of time. Not explained was the continual creation of new species. The problem of the origin of species had become an important research problem that an ambitious young naturalist—like Charles Darwin—could pay some attention to.[5] His theory of evolution answered a scientific question that in the 1830s was being called "the mystery of mysteries."[6]

EVOLUTION: DARWIN, WALLACE, AND SPENCER

Charles Darwin (1809–1882) appropriately holds a position of central importance in this essay. As can be inferred from what has already been said, it was not that Darwin's ideas were the only ones of significance for the nineteenth century. Far from it. However, his theory of evolution, formulated in the context of nineteenth-century ideas and issues, provided by the end of the century a major part of man's understanding of nature and of himself. Moreover, Darwin's religious ideas evolved through a pattern repeated in the minds of many Victorian intellectuals.

Darwin's famous voyage around the world in the early 1830s on H.M.S. *Beagle* convinced him that evolution had taken place. He apparently began the voyage in general agreement with Cuvier's view of geological catastrophes and fixed species. During the voyage, he read Lyell and became a uniformitarian, accepting Lyell's biological and geological views. He interpreted, for example, the results he observed of an earthquake as support for Lyell's theory of the formation of mountains. It was in 1837, the year after the voyage, that Darwin decided evolution had occurred.

With an eye toward geographical distribution of organisms, Darwin saw a striking commingling of diversity and similarity. Along the long east coast of South America, he recorded species giving way to closely allied species as he traveled from north to south. In Argentina, he found fossil remains of a large extinct animal similar to the modern armadillo that lived in the area. Other naturalists had made comparable observations, but Darwin responded to them differently. To Darwin they suggested continuous connections between past life and present. Rather than obliteration followed by entirely new formation of life, gradual transformation from the extinct form to the modern form seemed to Darwin more reasonable. In the case of the armadillo, such change would have accompanied environmental changes over a long period in a limited geographical area. In the case of South America's east coast, present forms seemed to be the result of diverse modifications over a long period of some one past species, the different modifications being associated with the different environments existing over this extensive geographical area. Observations on the Galapagos Islands supported his conclusions. Here Darwin found that each island had its own set of species. The familiar finches, for example, differed from island to island; moreover, all resembled species on the mainland. Apparently, once the species had been introduced from the mainland into the islands, their effective separation from one another had allowed each island population to evolve in its own way.

Not until Darwin had been back in England for two years did he formulate what appeared to be a plausible explanation of why evolution would occur. The clue, he tells us, came from his reading of Thomas Malthus's *On Population*, a widely read book that discussed issues of common interest to the naturalist,

the theologian, and the social theorist.[7] Malthus's warning that human populations tended to increase much more rapidly than the available supply of food
impressed on Darwin the immensity of the pressures exerted by animal populations on their rather fixed food supplies. A few simple calculations convinced
Darwin of the enormous constraints limiting animal populations and of the
large numbers that died compared to those that survived. Population pressure,
therefore, was the driving force of evolution by natural selection.

According to Darwin, random changes always occur between parents and
offspring. In the intense competition for survival, a few of these changes will
be beneficial, allowing the offspring possessing them to live longer on average
and to produce more offspring of their own than would otherwise have been
possible. Since these offspring also tend to possess the initial beneficial change,
the stage is set for future beneficial changes to be added to the first, accumulating gradually into a large enough change that we can say that one
species has evolved into another. In this way, for example, a population of
animals could change to meet the demands of a gradually changing environment. Or, portions of the population could change in different ways to fill
different niches in existing environments.

Finally, we should mention Darwin's use of what he called "artificial"
selection and his conception of the direction of evolution. Animal breeders, he
observed, were quite talented at producing the kind of animal they desired by
selecting which animals would mate. Their successes helped convince Darwin
of the great malleability of animal forms. Pigeon breeders, for example, had
been able to produce a number of different types of pigeons from an initial
kind. Hence, existing types of pigeons had been developed, not from each
other, but from a common ancestor in a kind of branching development. In
nature, too, Darwin thought, the course of evolution was like a tree, not a lad-
der. Thus, evolution did not proceed *toward* man. Man was simply the latest
result on one of several branches. Also, evolution along any particular branch
did not necessarily *progress* toward that branch's latest result, for some changes
went in different directions.

Consequently, Darwin's theory of evolution by natural selection was a
brutal process in which death, not life, was the rule. And the record of species
that lived reflected neither necessary progress nor unidirectional development
toward man. Indeed, in this random interaction between life and its environment, the appearance of man was by no means a foregone conclusion.

In religion, Darwin followed that common Victorian route from conventional Christianity to agnosticism.[8] Though an avid reader of Paley's works as
a Cambridge undergraduate, Darwin had abandoned traditional Christianity
by the late 1830s. The transformation was not entirely due to scientific
developments. Instead, he complained of such things as dubious history in the
Bible, unacceptable actions of the Old Testament God, and repugnant doctrines of Christianity. At the time of the publication of the *Origin of Species*
in 1859, Darwin considered himself a theist, basing his view on an internal conviction that such a wondrous universe could not be the result of pure chance.
Because of this view, he later thought, he tended in the *Origin* to personify

nature, making natural selection more akin to artificial selection than he meant. He rejected such personification when it was brought to his attention, and, indeed, in the years after 1859 the implications of evolution undermined his theism, leaving agnosticism in its place. The *appearance* of design in living things, which had fooled Paley and others, was actually the result of unde-signed random chance. The evolutionary process could not be that of an in-telligent, kind God because the process required such intense suffering from so many. The reality that man observed fit with the concept of natural selection but not with the notion of divine design.

A. R. Wallace (1823–1913) and Herbert Spencer (1820–1903) both arrived at theories of biological evolution independently of Darwin, and they did not always agree with him. Wallace fashioned a theory almost identical to Darwin's during the 1850s, and their ideas were published simultaneously in 1858, the year before Darwin's much more substantial *Origin* appeared. However, Wallace came to disagree with Darwin on certain points. Their main disagree-ment concerned the adequacy of evolution to explain man's origin. Although Wallace had rejected Christianity in his youth, psychical research (spiritualism) convinced him of the existence of some overseeing intelligent being. He thought that the intervention of such a supernatural being was necessary to bridge the gap between animals and mankind. He also thought that such an intervention had bridged evolutionary gaps between matter and life and be-tween plants and animals and that it was now guiding the progressive evolution of human society.

Spencer had accepted Lamarckian evolution upon reading Lyell's account of it, long before he knew anything of Darwin's ideas. He went on to formulate an all encompassing evolutionary philosophy. It included development from matter to life to man to society and emphasized the importance of unrestrained capitalism for the continued evolutionary progress of human society. Spencer placed such conclusions in his scientific, or "knowable" realm, which he dis-tinguished from the realm of the "Unknowable." Since God (if one existed) fell in the Unknowable, Spencer regarded himself as an agnostic. Although he thought it impossible to know anything about God's existence or attributes and insisted that evolution did not proceed toward a predetermined goal, Spencer did think evolutionary developments were the effect of the Unknowable's influence in the observable world. Interpreting Darwin's work from his own Lamarckian viewpoint, Spencer thought Darwin had provided major support for his own philosophical system. For his part, Darwin—though he too thought evolutionary concepts applied in some ways to society—found Spencer's biological and societal arguments far too speculative, with meager grounding in empirical evidence. Spencer attained great popularity, nevertheless, as some Christians, for example, felt justified in replacing Spencer's Unknowable with their God, concluding that he for his purposes directed the grand evolutionary process described by Spencer.

Hence, biological evolution was compatible with a number of more general viewpoints. Also, though Darwin's writings provided the principal scientific support for biological evolution, he was not the only spokesman for it. In fact, when late Victorians said they accepted evolution, they often did not have Darwin's particular version in mind.

THERMODYNAMICS: KELVIN

Regarded by Lord Kelvin (1824–1907) as "the greatest reform that physical science has experienced since the days of Newton,"[9] the science of thermodynamics dated from the same decade as Darwin's *Origin of Species*.[10] In a complex way, several European scientists contributed over a period of many years to the development of the first and second laws of thermodynamics. It will serve our purposes, however, to focus on the explanations of the laws given by Kelvin, the major British contributor to the origin of thermodynamics. Afterward, we can ask what connections there were between the two powerful Victorian sciences of thermodynamics and evolutionary biology.

The first law of thermodynamics dealt with the "conservation" of energy; the second with what Kelvin called the "dissipation" of energy. According to the first law, the total amount of energy in the world remained constant, as can be illustrated, for example, by a falling rock.[11] A rock held at rest above the earth possesses a certain amount of "potential" energy because of its displacement from the earth's surface. When the rock is released, the potential energy is converted into the "kinetic" energy of the rock's motion. Just before impact, the rock possesses no potential energy and its kinetic energy equals its original potential energy. Upon impact, the rock's kinetic energy is converted into energy of sound, which can be heard, and energy of heat, which raises the temperatures of the rock and the ground. The sound gradually diminishes, being converted to heat and raising the temperature of the air in the process. Hence, the rock's fall involves the conversion of a certain amount of potential energy into an equal amount of heat energy. This example also illustrates the second law, which maintains that energy continually dissipates or becomes less usable. Although the amounts of energy in the initial and final situations are equal, it is only in the initial situation that there is a rock having the potential of accomplishing something by its fall. Kelvin explained that one could similarly analyze other transformations of energy, say of chemical or electrical energy into heat and motion. Hence, as Kelvin declared in 1852, there was "a universal tendency in nature to the dissipation of mechanical energy."[12]

By the 1870s, establishment of the kinetic theory of gases allowed Kelvin to explain the *statistical* character of the second law. According to the kinetic theory, heat was the result of the motion of molecules of matter. The faster the motion of the molecules of a gas (or of any material object), the hotter it was. Kelvin discussed the connection between the second law and the kinetic theory by employing "Maxwell's demon," thought up by the British physicist James Clerk Maxwell but named by Kelvin. Imagine a quantity of hot air and a quantity of cold air introduced simultaneously into the opposite ends of a closed container. The moving molecules of the two quantities of air will interact, producing a container filled with air at a uniform temperature. Generally speaking, collisions between the "hot" and "cold" molecules will have slowed the first, accelerated the second. Molecules will still be moving at a variety of speeds, but their average speed will be the same throughout the container. In accordance with the second law, the nonstatic state with a temperature difference has dissipated into a static state of temperature uniformity.

Now imagine a very small creature who, in Kelvin's words, "can at pleasure stop, or strike, or push, or pull any single atom of matter, and so moderate its natural course of motion."[13] Place this "demon" between the hot and cold air, and he can keep the molecules from diffusing into a state of uniform temperature. Or, more strikingly, place him in a container with air at one temperature, and he can direct faster molecules in one direction and slower ones in the opposite, thus producing hot air at one end of the container, cold at the other. As Kelvin explained, "By operating selectively on individual atoms [Maxwell's demon] can reverse the natural dissipation of energy, can cause one half of a closed jar of air, or of a bar of iron, to become glowing hot and the other ice-cold. . . ."[14] The second law was a statistical law. The results produced by the imaginary demon could actually occur in nature, but it was almost infinitely improbable that they ever would because it would require vast numbers of *randomly* moving molecules to move in the same way at the same time. As Kelvin said, the " 'Dissipation of Energy' follows in nature from the fortuitous concourse of atoms."[15]

Kelvin also discussed various far-reaching issues to which thermodynamics related, or at least appeared to relate. For example, did the universal tendency to the dissipation of energy foretell an eventual static condition for the universe — a state in which all energy had dissipated to heat at the same temperature, thus precluding any further activity? Kelvin thought not. He argued that since one could not deny the *infinity* of the universe (which meant an infinite supply of energy), "science points rather to an endless progress, through an endless space. . . ."[16] Second was the question of whether or not the laws of thermodynamics governed the activity of living things. As early as the 1850s, Kelvin was convinced that life did not violate the first law (that is, could not create or destroy energy) and probably did not violate the second, although that was a possibility. No matter how completely living processes might be explained thermodynamically, however, Kelvin insisted that the existence of human consciousness and sense of free will placed the essence of life beyond physical principles, beyond scientific explanation. As a consequence, the origin of life required a creator. The origin of life *on earth* might not have involved a creative act, however, for Kelvin speculated that the earth's initial seeds of life could have been brought by meteors that had been in touch with life elsewhere in the universe.

The issue germane to Darwinian evolution was not one of these, but the problem of the age of the earth. Whatever the distant history or future of the whole universe and whatever the ultimate origin of life, was the earth itself old enough for evolution to have occurred once life was present on it? In 1865 Kelvin dropped a two-page bomb on uniformitarian geology: "The 'Doctrine of Uniformity' in Geology Briefly Refuted."[17] Here and elsewhere, he argued that the continually cooling earth must have altered enormously during its history and could not have existed for endless ages in more or less the same condition. Moreover, it could only have been habitable for a restricted period of time, on the order of 100 million years. Kelvin recognized that this restriction did not disprove the reality of evolution, but he thought it showed evolution could not have occurred by such a slow process as natural selection. And it was the randomness of natural selection that Kelvin disliked, because he thought it overlooked God's design, which was everywhere apparent. Kelvin's estimates of the earth's age were later overthrown by the discovery of radioactivity, but in the nineteenth century they looked valid and were generally accepted, at the expense of uniformitarian geology's vast time scale. Consequently, though Darwin formulated his ideas with the aid of uniformitarian geology, evolution gained wide support during a period when Kelvin's thermodynamics had undermined uniformitarianism.

STUDIES OF THE BIBLE: *Essays and Reviews*

Essays and Reviews marked a milestone in Anglican perceptions of the Bible. Published only a year after Darwin's *Origin of Species,* the book was little influenced by evolution. However, the book's seven Anglican authors (six of them clergymen) had absorbed the lessons of geological research, which they regarded as demonstrating the scientific inaccuracy of Genesis. Geology showed, they thought, that Genesis was not God's absolute word of truth for all times and places, but contained a large human element rooted in the specific context in which it was written. Moreover, German biblical scholars had been studying the Bible from the standpoint that it reflected the times in which it was written. Aware of such research well before 1860, British scholars knew more than they had publicly proclaimed. *Essays and Reviews* was meant finally to set the results of science and scholarship squarely before British readers. The authors pointed to the progress of human knowledge and to its consequence, that modern man knew more than ancient. Understanding of the Bible, they declared, should incorporate such improvements in knowledge. "The Christian religion is in a false position when all the tendencies of knowledge are opposed to it. Such a position cannot be long maintained, or can only end in the withdrawal of the educated classes from the influences of religion."[18]

By treating the Bible the same as any other ancient source, the contributors to *Essays and Reviews* hoped to discover what the *writers* of the Bible meant. They sought a straightforward account of the ancients' views, without reading into them later religious doctrine or modern scientific knowledge. "When interpreted like any other book," they were confident "the Bible will still remain unlike any other book."[19] The modern student of the Bible, they argued, "is not afraid that inquiries, which have for their object the truth, can ever be displeasing to the God of truth; or that the Word of God is in any such sense a word as to be hurt by investigations into its human origin and conception."[20]

What were their specific conclusions relating to Genesis? Historical and philological research indicated that Genesis and the other books traditionally believed to have been written by Moses had actually been accumulated from several sources over a substantial period of time. Research with Egyptian and other records stretched human history to at least a thousand years before the traditional biblical value of 4000 B.C. The Anglican authors reported that Genesis begins with two, quite different creation stories produced by different writers: "This is so philologically certain that it were useless to ignore it."[21] Moreover, the two stories differ irreconcilably from modern knowledge. In the first story, for example, the creation of light on the first day, *before* the creation of the sun on the fourth, represented an ancient misconception that the sun was not responsible for daylight because daylight preceded the rising of the sun and lasted after the sun set. Second, the vault of heaven created on the second day was "a permanent solid vault." The Hebrew word employed meant "something beaten out," like a metal plate — "It has been pretended that the word *rakia* may be translated *expanse,* so as merely to mean empty space. The context sufficiently rebuts this."[22] Third, the order of the appearance of plants and animals in the biblical days of creation is not the same as the order given in the geological record, so that one could not salvage the scientific accuracy of Genesis merely by interpreting the "days" as vast periods of time. Consequently, it was a mistake to think either that Genesis presented modern science in vague or disguised form or that modern science was wrong because it differed with Genesis. Mature thought recognized that the straightforward meaning of biblical writers was, in a scientific sense, quite wrong. Genesis would thus retain its proper "dignity and value" only "if we regard it as the speculation of some Hebrew Descartes or Newton, promulgated in all good faith as the best and most probable account that could be then given of God's universe."[23]

I have found that the authors of *Essays and Reviews* mentioned evolution three times. One suggested "that the supply of links which are at present wanting in the chain of animal life may lead to new conclusions respecting the origin of man," and declared it a "false policy" to "peril religion" by pitting revelation against such a possible finding.[24] The contributor responsible for analyzing the creation stories dismissed evolution as "an hypothesis not generally accepted by naturalists."[25] Only the writer discussing miracles supported evolution. Generally, he argued that belief in miracles, though typical of ancient times, was not possible in the modern era of scientific knowledge of the regularity of nature. Once taken as the major proof of Christianity, miracles now posed a major embarrassment. The answer was to accept Christianity because of the excellence of its message, not because of alleged miracles from long ago. Darwinian evolution entered the argument as one specific example of supposedly miraculous events yielding to nature's regularity with the advance of science.[26]

Essays and Reviews did not immediately carry the day. On the contrary, it engendered intense opposition and controversy. Not until the 1890s did its overall viewpoint gain wide acceptance, and that came with the assistance of a like-minded book, *Lux Mundi,* edited by the future Anglican bishop Charles Gore in 1889. *Lux Mundi* incorporated evolution more completely than *Essays* had, and Gore's own contribution extended the idea of antiquity's limited knowledge to Christ himself. Since Christ appeared to accept erroneous views of his day, Gore concluded that Christ, in becoming human, must have divested himself of divine omniscience. Gore's thesis did not gain wide consent, but as a leading topic of debate it signified the enormous changes wrought in Anglican views by 1900.

PHILOSOPHY: HUME, MILL, AND HUXLEY

Hume, Mill, and Huxley were more critical than the authors of *Essays and Reviews.* Hume lived in the eighteenth century but he is included here because his views were echoed in the influential writings of Mill and Huxley. Hume's views found much more favorable reception at the end of the nineteenth century than they had at the end of his own. All three criticized miracles and other aspects of Christianity, but the focus here is on their analyses of the design argument so closely allied with early nineteenth-century Christianity.

In his *Dialogues Concerning Natural Religion,* David Hume (1711–1776) brought the design argument under the scrutiny of his empirical philosophy. What is most important, he declared, is that the design argument is based on a false analogy. Consider the familiar exposition involving a watch, which states that just as we know the pieces of a watch were put together by a watchmaker, so also the intricately interlocking pieces of the universe must have been put together by an intelligent being. First, how do we know the watch *was* made by a watchmaker? The reason is clear. We have accumulated a good deal of experience of watches being made by watchmakers. Hence, if we find a watch, we reasonably infer that it, like the other watches, was made by a watchmaker. Second, why can we *not* argue by analogy to the universe as a whole? The reason is that because we have not observed other universes being constructed, we have no relevant accumulated experience on which to base such an inference. The case of the universe is different from that of the watch, because we know about only one universe.

Note that Hume does not say that we know that *no* god made the universe. His claim is that we have no way of knowing whether a god did make it or not. On the basis of available evidence, it is just as probable that the matter of the universe is all that exists and that it possesses some self-organizing principle responsible for the material universe as we observe it.

If one refuses to recognize the fatal flaw in the design argument and insists on the existence of an intelligent designer of the universe, then, Hume says, we should at least be aware of what the evidence does and does not allow us to infer. First, the universe may have been constructed by *several* gods. If we insist on inferring divine design from examples of human design and observe that numerous human-made items are constructed by many people working together, it is not possible to limit ourselves to only one god. Second, we certainly cannot infer a god with the properties — like omnipotence and beneficence — usually attributed by Christians to their god. The presence of tragic diseases and natural disasters precludes the existence of the supposed Christian god. Indeed, Hume suggested, perhaps our universe was made by a young and inexperienced god who did rather a bad job of it before going elsewhere to make better universes, having profited by his mistakes on us.

To the surprise of many, John Stuart Mill (1806–1873) turned out to be less skeptical than Hume. His posthumously published *Three Essays on Religion* (1874) revealed Mill to be more friendly toward Christianity than his earlier writings had indicated. Not that Christians, especially traditional Christians, could take much heart from what he had to say. Lukewarm toward the claims of biological evolution, he did think there was design in nature, mainly in living organisms. He declared, moreover, that the design argument provided the only possible indication of God's existence and attributes, but that, in accordance with the latter part of Hume's discussion, the design argument established the existence of a limited god, with matter probably existing independently from him. The existence of such an imperfect god at least lent a certain plausibility to the notion that the imperfect Scriptures were his word, and one could therefore "hope" — but no more — that the Bible might be this god's revelation. Chiefly, Mill emphasized the *usefulness,* rather than the truth, of Christianity in providing a moral code and a moral man for people to emulate.

Unlike Mill and the authors of *Essays* who paid little attention to evolution, the distinguished biologist T. H. Huxley (1825–1895) was Darwin's loudest supporter. For Huxley, Darwinian evolution explained the intricate design in animals that had so amazed Paley, and even Mill. It was Huxley who invented the word *agnostic* to describe his own philosophical position that neither the design argument nor any other argument could demonstrate the existence or nonexistence of a god. Exceedingly critical of biblical evidence, he was especially hard on Jesus' concept of evil demons as the cause of human ailments. When Jesus spoke of driving demons out of a man into a herd of swine, he disclosed himself as a man of his times, not a divinity with perfect knowledge. Posing what he regarded as an insoluble dilemma for Christianity, Huxley declared,

> Either Jesus said what he is reported to have said, or he did not. In the former case, it is inevitable that his authority on matters connected with the "unseen world" should be roughly shaken; in the latter, the blow falls upon the authority of the synoptic Gospels. If their report on a matter of such stupendous and far-reaching practical import as this is untrustworthy, how can we be sure of its trustworthiness in other cases?[27]

END OF THE CENTURY

What then was the late-century reaction to all these developments? Although they were not entirely separate from one another, we can speak of a scientific reaction and a religious reaction.

There were solid scientific objections to Darwin's theory. If Darwin's small beneficial changes were few in number, would they not almost certainly be swamped out of existence as the "improved" offspring mated with ordinary animals? If parental characteristics blended to produce those of the offspring, it was difficult to see how the beneficial changes could ever accumulate. Even without the problems posed by the blending of parental traits, Kelvin's calculations gave an age of the earth much younger than Darwin thought was required. Darwin's response to such concerns generally coincided with that of the late-Victorian scientific community. Although evolution remained the best available explanation of the fossil record and the geographical distribution of plants and animals, the cause and rate of evolutionary change were still puzzles. Faced with strong criticism, Darwin supplemented natural selection with Lamarckian ideas of the inheritance of acquired characteristics in order to make evolution a speedier process. In similar fashion, British scientists accepted the reality of evolution fairly readily but were still debating its causes when the century closed.

Religious reaction apparently took its cue largely from scientific considerations. While scientists argued, clergymen naturally looked askance at evolution. In his famous debate in 1860 with Huxley, for example, Bishop Samuel Wilberforce was coached by the eminent comparative anatomist Richard Owen. Eventually the various nineteenth-century developments seem to have divided those who remained Christians into two main groups—conservative opponents of evolution and liberal supporters of a divinely ordained, goal-directed evolution of plants, animals, man, and human society.

The final decade of the nineteenth century, therefore, differed enormously from the first. Late Victorians regarded 100 million years as a great *limitation* of the earth's age. Biological consensus embraced evolution, although biologists would have liked to have had a more definitive understanding of its causes. It had been decades since any legitimate claim could be made for the Bible as a science textbook. Under the persuasion of Mill, Darwin, Huxley, and biblical scholars, traditional Christianity had largely yielded to liberalism and agnosticism. Christians no longer appealed to Paley's design argument so frequently or confidently as they once had. In contrast to their early-century counterparts, they tended to regard science and religion as rather separate entities. Within its domain science appeared autonomous. It seemed wise to avoid the past mistake of tying religion to conclusions that science might one day disprove.

EPILOGUE: CANVASSING THE ROYAL SOCIETY

In 1932 C. L. Drawbridge, an 1891 graduate of Cambridge University and longtime secretary of the Christian Evidence Society, published *The Religion of Scientists*. On behalf of his society, he had sent a six-part questionnaire on science and religion to the fellows of the Royal Society of London; his book reported on the 200 responses he received. Standing midway between Darwin's death in 1882 and our own day, Drawbridge's little book nicely documents the continuation of certain trends already under way by 1900.

Evolution, for example, was assumed to be correct, and the questionnaire asked, "Is it your opinion that belief in evolution is compatible with belief in a Creator?" Although the fellows answered overwhelmingly yes, that did not necessarily mean they believed in the Christian God, as Drawbridge pointed out. If anything, the scientists' answers illustrated another trend—a decline in traditional Christian views among scientists. The most specific question in this respect asked, "Do you believe that the personalities of men and women exist after the death of their bodies?" To this only 47 of the 200 said yes, the rest either answering no, saying they could not decide, or choosing not to answer the question. Moreover, the explanatory comments accompanying many affirmative replies tended to be guarded statements like, "I know of no scientific evidence that they do, but think this belief to be the best working hypothesis for life."[28] Running through the book was the fundamental concept that science and religion occupy separate domains, which was Drawbridge's own view: "Natural science and theology speak different languages."[29] Summarizing the scientists' written responses, Drawbridge noted that "a large number of the Fellows of the Royal Society say that natural science has nothing to do with religious beliefs,"[30] and he was able to reproduce several statements of the form, "I think that science (in the sense of physical or natural science) has nothing to say for or against [the existence of a personal God]."[31] The scientists were far from unanimous in their views, but in the balance of their opinion we find a group radically different from what a similar group would have been a century or so before.

Nineteenth-century science and scholarship produced a conceptual revolution of the first magnitude, and we—like the fellows of the Royal Society half a century ago—live in its wake. Much has happened since 1900, of course. Physicists, biologists, anthropologists, and biochemists have extended thermodynamics and evolution in significant ways. Philosophers of science have debated the exact grounds of scientific knowledge, and biblical scholars have deepened our knowledge of the Bible. Yet these developments have occurred within an overall frame of understanding, which, as we have seen, was essentially established by 1900.

History, however, is not deterministic. We have traced changes in nineteenth-century thought, but such changes did not have to happen in exactly the way they did or when they did. More important, tracing the historical development of modern views does not *justify* those views. Knowing how we got here helps us understand why we are here but does not require us to stand still. Indeed, one of the surest lessons of history is that, in one way or another, we will not stand still. Hence, although this chapter has explained how the age of Darwin helped shape ours, one must look to later chapters for justification of current views and indication of possible areas for future change.

FOR FURTHER READING

ALLEN, GARLAND E. *Life Science in the Twentieth Century*. Cambridge: Cambridge University Press, 1978.
    Begins with the influence of Darwin on late nineteenth-century biology but concentrates on the main themes in biology in the twentieth, including developments in evolutionary theory.

BURCHFIELD, JOE D. *Lord Kelvin and the Age of the Earth*. New York: Science History Publications, 1975.
Discusses Kelvin's various calculations of the age of the earth, their impact on geological thought, and their eventual downfall.

CHADWICK, OWEN. *The Victorian Church*. 2 vols. New York: Oxford University Press, 1966–1970.
Examines many aspects of the Victorian church, including the controversies involving science and religion and the reception of *Essays and Reviews* and *Lux Mundi*.

COLEMAN, WILLIAM. *Biology in the Nineteenth Century: Problems of Form, Function, and Transformation*. New York: John Wiley, 1971.
Treats the chief aspects of nineteenth-century biology, including evolution.

EISELEY, LOREN. *Darwin's Century: Evolution and the Men Who Discovered It*. Garden City, N.Y.: Doubleday, 1961.
Devotes two chapters to Darwin and several to the background of his work and its reception.

GILLISPIE, CHARLES COULSTON. *Genesis and Geology: A Study in the Relations of Scientific Thought, Natural Theology, and Social Opinion in Great Britain, 1790–1850*. New York: Harper & Row, 1951.
Examines the maturing science of geology and its connections with religion in the decades before Darwin's *Origin of Species* was published.

GREENE, JOHN C. *The Death of Adam: Evolution and Its Impact on Western Thought*. Ames: Iowa State University Press, 1959.
Traces the revolutionary change from the static creationism of Newton's time to the evolutionary theories of Darwin's time.

MOORE, JAMES R. *The Post-Darwinian Controversies: A Study of the Protestant Struggle to Come to Terms with Darwin in Great Britain and America, 1870–1900*. Cambridge: Cambridge University Press, 1979.
Includes a critique of misleading "warfare" interpretations of the period, a concise summary of different nineteenth-century versions of evolutionary theory, and an extensive bibliography.

PORTER, ROY. *The Making of Geology: Earth Science in Britain, 1660–1815*. Cambridge: Cambridge University Press, 1977.
Examines the social context of developing British thought about the earth to show how the science of geology is a social artifact invented in a particular time and place.

REARDON, BERNARD M. G. *From Coleridge to Gore: A Century of Religious Thought in Britain*. London: Longman, 1971.
Surveys nineteenth-century British theology, including the content of *Essays and Reviews* and *Lux Mundi*.

RUDWICK, MARTIN J. S. *The Meaning of Fossils: Episodes in the History of Palaeontology*. London: Macdonald, 1972.
Discusses the growing knowledge of fossils and the fossil record, concluding with the complex relationship of such knowledge to Darwin's ideas.

RUSE, MICHAEL. *The Darwinian Revolution: Science Red in Tooth and Claw*. Chicago: Chicago University Press, 1979.
Analyzes the character of the Darwinian revolution, including its philosophical, theological, and political aspects.

STONE, IRVING. *The Origin: A Biographical Novel of Charles Darwin*. Garden City, N.Y.: Doubleday, 1980.
Reconstructs Darwin's life in great detail.

TURNER, FRANK MILLER. *Between Science and Religion: The Reaction to Scientific Naturalism in Late Victorian England*. New Haven, Conn.: Yale University Press, 1974.
Examines the thought of six intellectuals, including A. R. Wallace, who rejected Christianity but sought other ways of understanding religious issues, such as God's existence and mankind's immortality.

[2]

A Brief Critical Analysis
of Scientific Creationism

WARREN D. DOLPHIN

TOO often, scientists fail to take the time to listen to what creationists are saying and to understand the ramifications of their ideas. In this chapter we turn our attention to that failing by examining three aspects of scientific creationism that may help to clarify the movement for our readers. It is important to know, first, something of the history of scientific creationism and to understand why this apparent misnomer came into use.[1] Next, we should consider the evolution model alongside scientific creationism in order to pinpoint the actual areas of controversy. Finally, we must consider why the controversy has continued into the present and understand why it will no doubt continue into the future.

THE CREATIONISM MOVEMENT

Although creationists and creationism have been around from long before the time of Darwin, we need concern ourselves only with the developments after the publication of the *Origin of Species* in 1859. That time marks the beginning of the controversy between the scientific theory of evolution and religious groups that stressed a fundamental, inerrant interpretation of the Bible.

This controversy continues today in different guises. In the celebrated 1925 *Scopes* v. *Tennessee* trial the focus of the controversy shifted from abstract philosophical differences to the pragmatic issue of what should be taught in the public school science class. The verdict supported the Tennessee law prohibiting the teaching of evolution, but the trial is viewed as a popular victory for evolution. The effects of this trial on public education were negative, however, because most schools avoided teaching evolution and publishers produced texts that scarcely covered the topic. For thirty to forty years following the Scopes trial the creation-evolution controversy was not a high-priority issue.

The issue lay dormant until the 1950s when there was a growing realization that science education was weak in public schools throughout the United States. In the 1960s, a group of biologists under the auspices of the American Institute for Biological Sciences received a grant from the National Science Foundation to revamp the high school biology curriculum. This Biological Sciences Curriculum Study (BSCS) group produced a series of biology textbooks that used evolution as the major unifying theme. These materials were widely adopted and it is estimated that by 1970 over half the students taking high school biology were studying this branch of science via these materials.

The reappearance of evolution in high school classes brought a strong reaction from creationist groups, who argued before school boards and state legislatures that evolution should be banned from the classroom. Many states passed laws or attempted to enforce laws already on the books to effect such a ban. An Arkansas teacher, Susan Epperson, filed suit against the state's antievolution law, fearing prosecution under the law if she used the BSCS biology books that her school district had adopted. In 1968 the U.S. Supreme Court ruled that the state law banning evolution was unconstitutional because it favored a particular religious viewpoint over others.

As a result of this court case, creationists abandoned the position that evolution should not be taught. They adopted a new line, which demanded that both biblical creationism and evolution be taught equally in science classrooms as competing and mutually exclusive explanations of origins. Tennessee, for example, passed a law in 1973 requiring equal time for the Genesis account of creation whenever evolution was taught. The National Association of Biology Teachers challenged the law and in 1975 obtained a federal court judgment from the U.S. Court of Appeals that ruled the Tennessee law was unconstitutional in that it established a preference for teaching the biblical account of creationism over the theory of evolution.

Faced with two federal court rulings that blocked their efforts to ban evolution, or at least to teach biblical creationism, the creationists decided to regroup. Creationists stated that including only evolution in the curriculum as an explanation of origins was unconstitutional because it was the viewpoint of a religion they called secular humanism. They reasoned that Christian children, as well as atheistic secular children, have equal rights under the law in public schools. Since the U.S. Supreme Court had ruled that it was unconstitutional to teach about God in the public schools, then it was equally unconstitutional to teach the absence of God and, therefore, a way was sought to neutralize the teaching of evolution.

A new strategy then began to emerge in which creationism was to be considered principally as a scientific concept that happened to agree with the Bible. Creationism would no longer be presented as a religious doctrine, and thus the effects of previous court rulings could be circumvented or even used against evolution. Creationists said that since evolution is a theory that has not been proved, it requires a great deal of faith to be accepted. Faith is a strong component of religion; hence evolution is a religion. Therefore, since creationism and evolution both involve science as well as faith, they should be given equal time in the classroom.

This strategy was used before the California State Board of Education in the late 1960s to try to influence the contents of the state-approved syllabus for science. The then new scientific creationism was explained without reference to the Bible and was based on argumentation that pointed to the alleged impossibility of evolution and alleged faulty interpretation of geological, paleontological, and biological evidence supporting it. Several creationists with advanced degrees in science supported this effort, which led the California Board of Education to adopt a resolution leading to a modification of the syllabus to include a passage stating that the scientific evidence concerning the origin of life implies at least two and perhaps several possible explanatory theories. The roots of many scientific creationist organizations that are politically active today can be traced directly to this successful attempt to gain equal time for the creationist's viewpoint.

CREATIONIST ORGANIZATIONS

Creationist organizations include the Creation Research Society, the Institute for Creation Research, the Creation Science Research Center, Bible-Science Radio, the Bible Science Association, and Christian Heritage College. Many of these organizations are based in San Diego and appear to share a common leadership. They actively promote scientific creationism at the local and state levels throughout the nation. Paul Ellwanger heads a national organization called Citizens against Federal Establishment of Evolutionary Dogma, which grew out of the Citizens for Fairness in Education, a group advocating model legislation for states and calling for equal time for creationism and evolution. The legislation would affect the research funding programs of the National Science Foundation and the public education programs of the National Park Service and Smithsonian Institution. Some of these groups are apparently forming coalitions with other groups such as the Moral Majority and the Pro-Family Forum, which promote creationist materials developed by the San Diego groups. The increasing frequency of attacks on science and evolution is not due to chance. The creationists, now called scientific creationists, have organized to demand that their viewpoint be expressed. The remainder of this chapter addresses the question of whether these demands are "scientifically" worthy.

Since the Creation Research Society was among the first of the organizations to be formed and remains active today, it will be examined first to determine the active membership and their motivation for organizing. In the 1940s the American Scientific Affiliation (ASA) was formed by scientists whose objectives were to relate science disciplines to Christian religion. This group is discussed in greater detail later in the chapter. In 1963 members of the ASA who were fundamentalist Christians objected to the soft position that ASA espoused on the evolution issues. Ten members under the leadership of Walter Lammerts and William Tinkle split from the ASA to form the Creation Research Society, based in Ann Arbor, Michigan. This group—referred to as the team of ten by creationists—included Henry Morris and Duane Gish, who are very active in the scientific creationism movement today. According to materials printed in each issue of the society's journal, *Creation Research Society Quarterly,* the society is only a research and publications organization that does not hold meetings or engage in promotional activities. Its members purportedly represent the researchers on creation, and the society maintains a fund to assist in research projects. There are four classes of membership: (1) voting members, who hold at least a master's degree in some area of science or techology; (2) sustaining members, who are people interested in the work of the society; (3) student members; and (4) institutional subscribers. All members in the first three classes pay dues and must sign an application form subscribing to the following statements:

> 1. The Bible is the written Word of God, and because we believe it to be inspired thruout, all of its assertions are historically and scientifically true in all of the original autographs. To the student of nature, this means that the account of origins in Genesis is a factual presentation of simple historical truths.
> 2. All basic types of living things, including man, were made by direct creative acts of God during Creation Week as described in Genesis. Whatever biological changes have occurred since Creation have accomplished only changes within the original created kinds.
> 3. The great Flood described in Genesis, commonly referred to as the Noachian Deluge, was a historical event, worldwide in its extent and effect.
> 4. Finally, we are an organization of Christian men of science, who accept Jesus Christ as our Lord and Savior. The account of the special creation of Adam and Eve as one man and one woman, and their subsequent Fall into sin, is the basis for our belief in the necessity of a Savior for all mankind. . . .

The society is therefore committed to the biblical account(s) of the creation of the universe, earth, and living organisms. Furthermore, the society's statement of belief says that "we propose to re-evaluate science from this viewpoint."

In general, the articles in a society's journal represent the most current scholarship of that research community. It would be instructive for any person interested in the creation-evolution debate to obtain copies of the *Creation Research Society Quarterly* to see what scientific creation researchers are working on. When not attacking evolutionists, they are generally concerned with establishing the source and disappearance of the water of the Flood; discovering how the ark was constructed and was able to contain all the animals; and reconciling the recent findings of science in many fields with the Scriptures. A few titles from recent issues, with summaries, are listed below:

1. "Creationism and Continental Drift" — Purports that energy calculations that show continental drift can be explained either by intervention of God or by expansion of the earth in such a way as not to involve viscous forces.

2. "Is the Earth's Core Water?" — Proposes that the earth's core consists of water under great heat and pressure and concludes that the Bible supports such a hypothesis.

3. "Insect Flight: Testimony to Creation" — Views the great variety of flight mechanisms as evidence of the creator's skill and of his giving insects the means to make them fit for the life-style assigned them in his plan.

4. "The Hebrew Flood Even More Devastating Than the English Translation Depicts"—Explains the usage of special Hebrew words that appear in Genesis.

5. "The Necessity of the Canopies"—Water for the Flood came from a vapor canopy that surrounded the earth like the one found on Venus by space probes. The presence of the vapor canopy explains the tropical climate before and after the Flood, for example, the presence of mammoths in the Arctic. When the canopy dissipated, the Ice Age started.

6. "Thoughts on the Structure of the Ark"—Describes how the ark could have been made from balsa logs.

7. "Magnetic Fields in Medicine: Bone Repair"—Discusses the therapeutic effects of magnetic fields on bone fracture repair and the implications in explaining the life spans of the biblical partriarchs.

8. "Punctuated Equilibrium and Macro-Micromutation Controversy"—Reviews the controversy and urges creationists to be aware of these arguments in the scientific literature so they can collect arguments against evolution and hence, by default, in favor of creationism.

The subscriptions of belief and the types of papers in the society's journal make clear the questionable quality of the scientific scholarship among scientific creationists.

In 1970, Henry Morris helped found Christian Heritage College, at which all studies and curricula are organized around the concept of biblical creationism. As of 1981, the college offered B.A. and B.S. degrees in seventeen undergraduate majors and was a recognized candidate for accreditation by the Western Association of Schools and Colleges. An M.S. degree in science and science education was introduced in 1981. Originally, all classes were taught in the buildings of the Scott Memorial Baptist Church in El Cajon, California. Now the college is situated on a thirty-acre campus, has about 500 students, and 15 full-time staff plus 56 part-time staff. The college catalog states that the entire faculty supports the inerrancy of Holy Scripture, especially in relation to four major areas: (1) special creation of the world in six days, (2) biblical account of entrance of sin into the world, (3) the Noachian Flood, and (4) origins of nations and languages at the Tower of Babel.

The Creation Science Research Center (CRSC) was associated with the college when it was founded. In 1972 the center became an independent organization under the direction of Kelly Seagraves, and an Institute for Creation Research (ICR) was started at the college under Henry Morris's leadership. CRSC has become a political action organization, especially in California, while ICR has become an information center publishing creationism books and brochures. About twenty Ph.D.s are involved in the institute as staff or consultants. The annual budget is approximately $650,000. The stated purposes are to promote "research, writing, and educational ministries in the field of scientific Biblical creationism." The institute is currently supporting research studies to find the ark on Mount Ararat and to describe the physical evidence for a young (approximately ten thousand years old) earth. The goal of the writing ministries is to produce materials about scientific creationism for all grade levels within the framework of a truly biblical world view. Materials for both public and Christian schools are produced.

The most visible activity of the Institute for Creation Research is its publication of pamphlets in its "Impact Series," which are considered "vital articles on science/creation." A periodical called *Acts and Facts* is also published. Henry Morris, Duane Gish (a biochemist), Richard Bliss (a science educator), and Wendell Bird (a lawyer) are often authors in this series. The institute evidently carries out the promotional activities for creationism that are not in the purview of the Creation Research Society. A sampling of the approximately 100 titles in the "Impact Series" are:

1. "Resolution for a Balanced Presentation of Evolution and Creationism" — This pamphlet presents a model resolution that is intended for use by citizen groups when petitioning local school boards.

2. "A Two-Model Approach to Origins: A Curriculum Imperative" — Lays out a framework for teaching creation-evolution in the public schools.

3. "Thermodynamics and the Origin of Life, Parts I and II" — Articles written in response to suggestions that Ilya Prigogine, 1977 Nobel laureate in chemistry, had solved the problem of how the second law of thermodynamics could apply to evolution. Attempts to show the impossibility of his application.

Leaders of the Institute for Creation Research, especially H. M. Morris, publish books through a San Diego publisher called Creation-Life/Master Books. Several but not all of the books published are listed in the annotated bibliography at the end of this chapter. In addition, the publisher produces books on standard fundamentalist religious topics — for example, *The Bible Has the Answer, That You Might Believe,* and *The Unhappy Gays.* Various filmstrips and slide sets with audiocassettes describing creationism are also available. The creationist science textbook *Biology: A Search for Order in Complexity* by J. N. Moore and H. Slusher, written for use as a complete text in high school classes, is available from this source. Standard biological topics are covered, except for phylogeny and phylogenetic groupings. Both creation and evolution are presented as alternatives, but creationism is put forth as the superior explanation. The Indiana Superior Court (1977) found the clear purpose of this book to be the promotion of fundamentalist Christian doctrine in the public schools, and the court found the assertion that the text presented a clear and balanced case to be a sham.

These groups — a research society, a college, an institute, and a publisher — are vocal proponents of the organized scientific creationism philosophy. Although they state that the case can be made without biblical reference, each group shows a relationship to a fundamentalist Christian belief while excluding reputable scientists and scholars who happen to be Jews, Moslems, atheists, or members of other religions. Scientific creationists are evidently Christian, believe in literalness of the Bible, and are committed to reinterpreting the data of science in order to show compatibility with Genesis. By contrast, scientific societies and organizations place no such restrictions on their members.

CREATIONISM AND EVOLUTIONARY THEORIES COMPARED

If the myriad publications distributed by the creationist organizations are perused, it is possible to develop a model describing scientific creationism that can be used as a basis for comparing creationism and evolution. It should be apparent that scientific creationists and mainstream scientists have very different views of the world. For the reader who would like to examine some literature describing the basis of the scientific creationism theory, two books cited at the end of the chapter are particularly instructive. *The Scientific Case for Creationism* provides a brief but useful introduction to this view, while *Scientific Creationism* contains a detailed account that is considered by creationists to be the single most comprehensive source. Readers should be sure to request the unabridged version of the latter book, for it contains the "Creation According to Scripture" chapter that has been left out of the public school edition in order to present the case without reference to the Bible. Its inclusion in the general edition is strong witness by the creationists to the relationship of scientific creationism and religious doctrine, despite what they might say.

The conflict between the creationist and evolutionist viewpoints involves many areas of science besides biology. Table 2.1, which was modified from the writings of H. M. Morris, the director of the Institute for Creation Research and past president of Christian Heritage College, shows that the controversy also extends into the fields of astronomy, geology, and anthropology. A significant point not shown in the table is that scientific creationists also disagree with the views of many Christian sects and most philosophers of science and of religion. Several other chapters in this book address these differences in depth.

TABLE 2.1. CATEGORIZED COMPARISON OF DIFFERENCES BETWEEN EVOLUTION AND CREATIONISM

	Basic predictions	
Category of differences	Evolution model	Creation model
Astronomy		
Structure of stars	Can change into other types	Never change
Structure of universe	Building up and breaking down	Breaking down only
Geology		
Types of rock formations	Different in different "ages"	Similar throughout "ages"
Age of earth	Extremely old (10^9 years)	Probably young (10^4 years)
Fossil record	Innumerable transitions	Systematic gaps
Biology		
Origin of life	Life initially originating from nonlife	Life only from life
Diversity of organisms	Continuum of organisms over time	Distinct kinds of organisms from beginning
Appearance of kinds of life	New species appearing over prolonged periods	No new kinds appearing
Mutations	Random (most harmful; some beneficial)	Harmful
Genetic variation	Can be a creative process	Limited to specific themes
Natural selection	Window to future	Conservative process preserving past
Anthropology		
Appearance of man	Ape-human intermediates	No ape-human intermediates
Origin of civilization	Slow and gradual	Simultaneous with man

SOURCE: Modified from H. M. Morris, *Scientific Creationism* (San Diego, Calif.: Creation-Life Publishers, 1974), p. 131.

A broad interpretation of general evolutionary theory would state that a scientific explanation for all components of the universe can be found in the innate processes of the universe itself. The universe, therefore, was developed by its own means into its present structure and will continue to develop in this way. The diversity of living organisms has resulted from processes innate to them and the interaction between them and their environment. These processes must include reproduction mechanisms and metabolism. Evolutionary theory neither implies nor denies the existence of a divine being.

Henry Morris has eloquently stated the basic creationist position, saying that:

> . . . the universe could not have originated and developed itself but, rather, that its fundamental cause must be transcendent to its present processes and structure. It does not in any sense deny the scientific validity of these phenomena, but merely says that they could not have originated themselves and that, therefore, they must be explained in terms of unique creative processes which functioned in the past but are no longer operative at present.[2]

The creation model postulates that supernatural intervention by a creator is necessary to account for the complexity of nature. This is the basic premise of what is known as the design argument, which reasons that complex functional relationships imply a design and that a design must have a designer. Although many creationists contend that the *evidence* for a designer exists independently and can be examined without reference to the Bible, Henry Morris reveals their true position in his statement: "It should be stressed as strongly as possible that it is only in the Bible that we can possibly obtain any information about the methods of creation, the order of creation, the duration of creation, or any other details of creation."[3] One must question how such a statement from the director of the premier research institute on creation could be misinterpreted by other scientific creationists who publicly claim that the case can be made independently of the Bible. The message is clear: the Bible is the record and the only source of information about creation.

It would seem impossible for a case to be made for scientific creationism without reference to the Bible, unless the old debater's tactic of diversion is used. Debaters in any controversy can claim that there are only two possible explanations of a problem, theirs and the opposition's. Then the opposition is vigorously attacked through simple questions requiring technical answers. These answers are ridiculed and then further questions are rapidly posed. When the issue becomes hopelessly entangled, the opposition's position is proclaimed an impossibility and the argument is won by default. Creationists have used this technique in debates at churches, on college campuses, and before public bodies. Very little is ever said about creation in such debates, but many questions are raised about evolution. Once the issues are obscured, the debater proposes a suggestion that in the context of confusion seems reasonable: Wouldn't it be fair to teach both models? The audiences, confused and emotionally involved, usually answer in the affirmative.

Having recognized the source of the confusion, we can now identify the basic issue that should be under debate: Is it fair to teach scientific creationism as a scientific theory? To answer this question, both the evolution and the scientific creation models must be examined in detail.

BIOLOGICAL EVOLUTION MODEL

The biologist divides evolution into two categories, micro- and macroevolution. The former refers to the concept of evolutionary change seen in terms of the processes of variation, natural selection, and adaptation. Ideas about the processes of microevolution are stated in the form of testable hypotheses that can be studied within a lifetime in the laboratory or in field situations. Macroevolution refers to the concept of change seen in terms of the broader picture of present biological diversity. The ideas in this category arise from the application of microevolutionary findings to explain such phenomena as extinction, fossils, and historical genetic relationships among organisms (phylogeny). In so doing, macroevolution concerns itself with physical phenomena such as sedimentation, continental drift, radioisotopic dating methods, and other geological topics. Inherent in the macroevolution concept is the idea of a changing planet with a changing flora and fauna in a changing universe. It should be emphasized that evolution refers to phenomena that occur in populations and is not a characteristic of individuals. Individuals do not evolve; they only age and die. Populations, however, evolve over several generations because of changes in the genetic structure of the population, including but not restricted to those changes due solely to chance.

Microevolution may best be described by example. If we start with some ancestral population—such as the cabbage moths of last summer or the *Australopithecus* humanoid—all modern studies of genetics tell us there will be variability in gene content among the sperm and the eggs. The variability results from biparental inheritance, chromosomal crossing over (recombination), and mutations. Whenever meiosis (cell division forming sperm and eggs) occurs, these variability-introducing events happen, so that one can state that no two eggs or sperm are ever genetically alike. If these variable eggs and sperm fuse during fertilization, the members of the next generation will differ in some way from each other and from their parents. Hence, the next generation shows variation in its characteristics.

Among the members of the new generation, some characteristics may make certain individuals more "fit" than others; that is, they will function well in the total environment they inhabit. This situation allows the operation of natural selection. Natural selection is only definable within the context of an organism, a trait, and an environmental component. The overall result of natural selection acting on the total characteristics of an individual, not just on one trait, is that some individuals produce more offspring than others. It does not act by killing the "unfit" and leaving the fit alone. If interpreted this way, the familiar phrase "survival of the fittest" is misleading. The process is such that when an organism produces a large number of offspring, more of its genes and gene combinations are passed to future generations than are those of the less fit. Failure to reproduce adequately means few, if any, genes of that individual reach future generations. Thus natural selection provides a "window" to the future for each generation. The less fit will be crowded out by the more fit over several generations until the adapted types constitute most, if not all, of the population.

Macroevolution theory starts with certain underlying findings. These are:

1. Earth came into existence approximately 4.6 billion years ago by interaction of matter according to physical laws.

2. Complex organic molecules were formed from interaction of inorganic molecules and simple organic molecules—for example, given proper conditions, CH_4, H_2CO, H_2O, NH_3, and CO_2 combined to form amino acids, which in turn combined to form proteins.

3. Complex molecules formed structural complexes that could be considered the forerunners of cells (protocells).

4. Given protocells that could metabolize and were self-replicating, natural selection would act on any variations. This concept represents the incorporation of the whole of microevolution into macroevolution.

5. Given these conditions and the action of uniform physical and biological processes for billions of years, there was an increase in organic diversity, which is recorded in fossil remains and the relatedness of groups of organisms alive today.

The biologist would refer to this historical development of diversity as a phylogeny and would represent it as in the tree diagram in Figure 2.1, which attempts to show that diversity is a function of time from the events of the origin through the present to the future. Diversity can be interpreted as an increased number of species. Originally there may have been more than one spontaneous appearance of living forms, but scientists *currently* think that only one form was successful. From this original form, which had the ability to reproduce, other species have evolved through the forces of selection. Some of these were successful and continued to exist, changing in accordance with directional or diversifying selection. Others were unsuccessful and died off (e = extinction) as the environment changed. Note that as we come to the present in Figure 2.1, some lines diverge and take new directions, whereas others remain constant over long periods. The model is based solely on the extrapolation of the findings of microevolution, back in time to explain fossils and forward to indicate future changes. This diagram illustrates the important concept that all organisms are related. This relatedness and change through time are among the most important contributions of biology to modern thought.

SCIENTIFIC CREATIONISM MODEL

If one examines the assertions of scientific creationists other than their diatribes alleging the impossibility of evolution, it is possible to discern certain basic principles on which they attempt to build their model. First and foremost, they believe that creation was by the hands of the creator. Consider some of the details of this creation:

1. Scientific creationists believe that the world was brought into existence

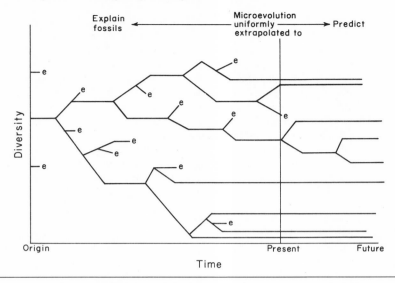

FIG. 2.1. Representation of biological diversity from the perspective of evolution. Diversity can be considered as the number of species and is plotted as a function of time in *billions* of years. If one looks at organisms today, their relatedness to others in the past and in the present is apparent. Hence, the notion of similar organisms belonging in similar groups is based not only on their characteristics but also on their sharing of common ancestors.

comparatively recently. A date of approximately 4000 B.C., which is based on biblical genealogies, is commonly cited. Other creationist authors cite an age of ten thousand years, which is based on calculations of the decay of the earth's magnetic field. Creationists believe that geologists, astronomers, and physicists fabricate and misinterpret data to arrive at an age of billions of years for the earth.

2. Scientific creationists think that although the creator created basic "kinds" of organisms, these were not exactly the same types as are alive today. To a certain extent, they accept the overwhelming data supporting microevolution and indicate that present forms may have diverged from the kinds originally created by the action of selection on genetic variability. This divergence is limited in their minds, however, and they insist that new kinds have never developed. The term *kind* as used by the creationist is not precisely defined and could be equivalent to a modern biologist's taxonomic species, genus, or family. This imprecision allows a certain flexibility in interpreting the fossil record of the horse, for example, because one can state that all the material that has been identified simply represents a record of variation from an originally created kind. It also allows scientific creationists to state that they accept modern genetic evidence on variation and selection and that they simply disagree on how microevolution can be projected back in time.

3. The above model is often modified by bringing in a biblical version based on events in Genesis. This submodel has the following components:

    a. The sequence of creative acts in Genesis is correct, including the creation of Adam and Eve, and creation actually occurred in six days. Creationists generally avoid stating which of the two Genesis stories is considered the correct sequence. How could one creation have happened in two ways?

    b. When the world was created, it was in a perfect state with minimum disorder (entropy). Starting in the Garden of Eden with Adam's fall, a universal process of decay and death came into being, which introduced disorder and imperfection. This process is equated with the increase in entropy described by the second law of thermodynamics, which creationists claim is not understood by most scientists.

    c. Some time after creation, the worldwide cataclysmic Noachian Flood changed the face of the earth and the nature and rate of most earth processes. The only organisms that survived the Flood were those on Noah's ark.

    d. A date of about 4000 B.C. is given for creation. Archbishop James Usher calculated this date in the seventeenth century on the basis of the genealogies in the Bible.

    e. The races and languages of men are traceable to the Tower of Babel.

4. The scientific creationists believe that once the creator finished creation, all the creating processes ceased and conserving processes were started to maintain the world and enable it to accomplish the purposes the creator had in mind when the project was started.

If a graph were made of the creationist concept of biological diversity over time, it would look like Figure 2.2. All "kinds" of organisms were created independently in the beginning by a supernatural creator and none has been created since, although some have become extinct. Not too long after the creation, the Noachian Flood occurred in which several "kinds" became extinct. The only "kinds" that survived were those on Noah's ark, and they gave rise to present-day forms, though further extinctions may have occurred. Creationists believe that in a few cases, microevolution could have caused variation in a basic kind, as in the canine where it gave rise to the dog, wolf, and fox, but these are not new kinds. They are still canines. In this model there are few if any relationships among currently living organisms other than those that were put there when creation occurred.

The scientific creationists believe that certain predictions can be derived from their model and that these provide proofs of its validity while disproving the evolution model. The laws of thermodynamics are thought to provide conceptual support for creation. They reason that because creation was complete and perfect, there is nothing to be added or subtracted from the present order of things in accordance with the first law of thermodynamics, which states that neither energy nor matter can be created or destroyed. The creationists also purport that the second law of thermodynamics supports creation by predicting a universal process of decay. Simplistically stated, the second law says that any

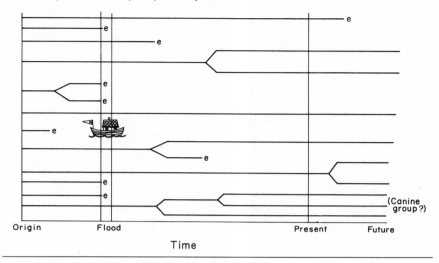

FIG. 2.2. Representation of biological diversity from the perspective of scientific creationism. Diversity can be considered as the number of kinds of organisms, and is plotted as a function of time in *thousands* of years. In contrast to the evolution model, the creationist model leads one to see little relatedness in present forms. Note also that in the past there were more kinds of organisms alive than there are today and that the number of kinds can only decrease in the future. This model is testable in a scientific sense. One need only prove that all organisms coexisted at some time in the past to confirm this model and discredit the evolution model.

time there are energy exchanges, a certain amount of energy will become unavailable to do work. "Entropy" is a measure of such unavailability, and it is universally acknowledged that "ordered" systems tend to degrade to less ordered systems while entropy tends to increase. Creationists point out that at creation everything was perfect (that is, ordered) but now things are in disarray, as we all know from common experience. This disarray is explained by the second law of thermodynamics. Evolution, creationists claim, is counter to the laws of thermodynamics because it suggests that the universe proceeds from the simple (disordered) to the complex (ordered) rather than in the opposite direction. What they completely overlook, of course, is that in biological systems order can be created from disorder at the expense of energy input from an external source, the sun. Their arguments are spurious at best and are based on a misleading simplistic approach to a complex, quantitative scientific concept. In brief, they have created a clever word game by introducing thermodynamics into their arguments.

When approaching the question of biological similarity within and diversity across groups of organisms, the creationists believe that their model has predictive powers. A designer, they suggest, would come up with similar designs for basic kinds whenever similar functions had to be carried out by organisms in similar environments. Diversity would be expected along with gaps in the fossil record because each type of organism has its own created purpose, which was part of the creator's plan. There is no doubt that in order to explain diversity, the creationists must appeal not to natural science but to supernatural intervention.

When the Flood component of the biblical model is considered, several more predictions can be derived. Mass extinction of organisms is simply a consequence of their not being on the ark. Sedimentary rocks were laid down during the brief period of a year or so that the flood waters covered the earth. Fossils represent organisms caught in silt deposited by the Flood. There are certain problems, however, in explaining the Flood scientifically. If a Flood were to cover the highest mountains on the earth (about six miles high), a fair amount of water would have to come from somewhere, and then it would have to drain somewhere when the waters receded. Also, how did all the animals (no plants mentioned) from the various continents reach the embarkation point and then find their way back after debarkation? How did Noah contain and feed them on the voyage? The fate of plants during this yearlong flood is not mentioned in Genesis, yet any model would have to provide a mechanism for their survival as well.

Another problem for the creationists is how to explain the fossil record in sedimentary rocks. It contains, in a general sense, simple organisms in the lowest strata and complex organisms in the upper strata. How could this happen in a flood, which would be expected to cause random mixing? H. M. Morris, who is a hydraulic engineer by training, has coauthored a book explaining this phenomenon. In it the authors argue that primitive organisms generally live in sediments and, therefore, would have been covered first since more complex organisms were able to swim. Gradually, depending on the degree of complexity, the complex forms succumbed and were buried. Looking at the fossil record, however, we find complex organisms as well as simple organisms at the top. If the densities of sediments are examined, layers of coarse, heavy granules are often found on top of layers of light, fine silts. These findings are inconsistent with Morris's theory but not with the accepted ideas of gradual deposition of sediments and fossils over prolonged periods.

The Nature of the Problem

Whenever fellow biologists discuss scientific creationism, they shake their heads in disbelief and ask how anyone can consider such ideas to be scientific. The answer lies in the creationists' tremendous commitment to religious belief and biblical inerrancy and their fixation on making the relationships of science conform to their simplistic notions of the Scriptures. W. T. Stace has presented an interesting thesis in his discussion of seventeenth-century conflicts between religion and science in *Religion and the Modern Mind*. Whether or not Stace's remarks accurately portray the history of science, they do seem to unveil the creationists' fears about what science has accomplished in the modern world. Stace thinks that the discoveries of science beginning in the 1600s psychologically increased the remoteness of God by pushing back divine intervention from the time of man's origins as described in Genesis to a remote time at the beginning of the universe:

> That God personally and directly created our ancestor Adam means that he was interested in man, and this gives a glimmer of warmth even to our present relations with him. Perhaps this God, though he does not act now, is still looking at us and thinking of us. But if the race of men is only a by-product of natural laws set going millions of years ago, then—even if God created the natural laws—what can man be to God, or God to man?[4]

He goes on to develop the idea of remoteness by examining the methods of science as follows:

> Science has everywhere substituted natural causes of phenomena for the supernatural causes in which men formerly believed. Since the seventeenth century it has become a fixed maxim of science that no supernatural causes are to be admitted in the universe; that even if a natural phenomenon is at present unexplained, it must never be taken as a case of divine action, but always a natural cause must be assumed to exist and must be sought for. This . . . is not a particular discovery of any particular science. . . . It is a basic assumption of science as a whole. . . . And it is this general scientific view of the world, and not any particular discovery, which has worked havoc with religion.[5]

The findings and methods of science, it is argued, result in an apparent remoteness between man and God. The scientific creationist, by means of apologetic arguments, seeks to reduce the distance by returning to fundamental religious concepts. In so doing, the creationist brings another specter into focus: moral disorder. There is no doubt that creationist literature combines concern for the supposed fallacies of evolutionary theory with concern for the decay of morality in today's world. Stace recognized this general trend in 1952 and put the argument in a form similar to the one outlined below:

FUNDAMENTALISM VERSUS SCIENCE

1. If morality is grounded in divine purpose, it is objective and the world has a single universal morality.
2. However, science has caused a loss of effective belief in divine purpose by inserting natural processes.
3. Hence morality no longer seems to be grounded in divine purpose.
4. Since values must be connected with purpose and if divine purposes are eliminated, the remaining alternative is human purpose.
5. It follows then that moral value could very well be subjective, existing only in the minds of men, and will cease to exist when they do.
6. Therefore, morals could be relative, not absolute, and the world did not have moral order built into it.

Henry M. Morris was evidently concerned with this type of problem when he wrote: "If man is a product of evolution, he is not a fallen creature in need of a Savior but is a rising creature capable of saving himself. . . . Evolution's gospel yields materialism, collectivism, anarchism, atheism, and despair in death."[6] The Jehovah's Witnesses, a fundamentalist religious group, have expressed a similar view:

> If man is created, then this implies he was created for a purpose, which in turn is suggestive of man's responsibility to his Maker. The desire to be free from such responsibility has turned many to evolution and atheism. . . . The prevalence of wickedness is an important reason why many turn to a belief of evolution.[7]

If the interpretations of the nature of the creation-evolution controversy presented in this chapter are accurate, the proof that scientific creationism is not science will not end the discord. Creationists are committed to a religious, not a scientific, cause. Scientific creationism appears to be a tactic conceived in the late 1960s to suggest that creationism could be taught in the classroom as science rather than religion. The tactic was successful in California and now it is being applied nationwide. Scientists and educators are rallying to indicate the absurdity of teaching what is obviously pseudoscience and religion. The issue may wither but it will not die. Religious and institutional commitments are too great.

In ending this chapter, I would caution the reader not to fall into the trap of believing that there are only two sides to the issue. Attempts have been made here to describe scientific creationism in terms of tactics, theories, and issues in order to throw some light on the controversy. You are also admonished not to accept evolution by default as the only explanation of origins. Scientific creationism as expressed by the San Diego groups was chosen because their antievolution position is extreme and they are politically active. There are, however, many others who support forms of creationism, presenting varying degrees of plausibility.

One such group is the American Scientific Affiliation, a society of 2,000 scientists founded in 1941 out of a concern for the relationship between science and the Christian faith. It is interesting to recall that the Creation Research Society was formed in 1963 by members of this group who thought that ASA had become soft in its opposition to evolution. The society publishes the *Journal of the American Scientific Affiliation*. Members endorse the following statement of faith:

1. The Holy Scriptures are the inspired word of God, the only unerring guide of faith and conduct.
2. Jesus Christ is the Son of God and through His Atonement is the one and only Mediator between God and man.
3. God is the Creator of the physical universe. Certain laws are discernible in the manner in which God upholds the universe. The scientific approach is capable of giving reliable information about the natural world.

The stated purposes of the society are "to explore any and every area relating

Christian faith and science" and to make known the results of such investigations for comment and criticism by the Christian community and the scientific community.[8] The society takes no official positions on controversial issues. There is no doubt that members of this group believe in creation, God, and natural laws.

Members of this group generally follow the theistic evolution theory. This theory suggests that a creator was involved in creating the universe; but since the creation, the universe has changed in a manner described by the natural laws we know today: thermodynamics, evolution, and so on. Theistic evolution accepts the geological evidence for an old earth and sees no reason to impose artificial barriers on the types of changes that can occur in the universe or in the earth's flora and fauna. Theistic evolutionists have been able to reconcile the logic of science with religious faith. It is unfortunate that the name scientific creationist has been tainted by extremists espousing pseudoscience and that it cannot be used to describe those men and women of many faiths who believe that the findings and methods of science do not exclude their belief in God.

EPILOGUE

According to Greek mythology, Odysseus was presented with a tactical problem during the Trojan War: how to get his men inside the bastions of Troy, which had successfully resisted previous besiegement. Troops that had formerly attacked the city directly were withdrawn and hidden a short distance away. In their place was left a large wooden horse sculpture in which were concealed several warriors. The Trojans accepted the wooden horse at face value as a peace offering and took it into their city. At night the warriors came out of the wooden horse, slew the guards at the gates of Troy, and opened the gates to their returning troops, who plundered the town.

We should learn from the lessons of history and not accept things as they appear on the surface. The core of an idea should be examined before it is accepted as being reasonable. Scientific creationism is to science as the horse was to Troy. Scientific creationism is not science. Its core is a narrow fundamentalist religious interpretation of Christianity, and its disciples are attempting to gain access to science so that they may undermine the discipline from within. Admission of scientific creationism to the classroom as science is not only a contradiction of terms but will lead to the wholesale destruction of a form of reasoning that serves us well.

FOR FURTHER READING

ANONYMOUS. *Did Man Get Here by Evolution or by Creation?* New York: Watch Tower Bible and Tract Society, 1967.
  A brief book based on the design argument and the Scriptures. Relates evils in the world today to belief in evolution.
Creation Research Society Quarterly. Ann Arbor, Mich.: Creation Research Society, 1964 continuing.

The principal journal on creation research. Information can be obtained by writing to the Secretary, Creation Research Society, 2717 Cranbrook Rd., Ann Arbor, Michigan 48104.

GISH, D. T. *Speculations and Experiments Related to the Origins of Life* (*A Critique*). San Diego: Institute for Creation Research, 1972.

GISH, D. T. *Evolution? The Fossils Say No!* 2nd ed. San Diego: Creation-Life Publishers, 1973.

Creationists consider this to be the most complete documentation of the absence of transitional forms in the fossil record, especially humanoid remains.

Journal of the American Scientific Affiliation. Ipswich, Mass.: American Scientific Affiliation, 1949 continuing.

A journal concerned with fostering communication on scientific and Christian issues. Information can be obtained by writing to American Scientific Affiliation, P.O. Box J, Ipswich, Massachusetts 01938.

MORRIS, H. M. *The Remarkable Birth of Planet Earth.* Minneapolis: Dimension Books, Bethany Fellowship, 1972.

Introductory treatment of origins covering early history of earth and mankind, delusion of evolution, the Flood, and historic confirmations of God's handiwork.

MORRIS, H. M. *Scientific Creationism.* Gen. ed. San Diego: Creation-Life Publishers, 1974.

Considered by creationists to be the best documented treatment of scientific creationism and is printed in general and public school editions with the latter lacking a chapter about the Scriptures.

MORRIS, H. M. *The Troubled Waters of Evolution.* San Diego: Creation-Life Publishers, 1975.

Describes the effects of evolutionary thought on education and society including the dangers inherent in application of evolutionary thinking to world problems. It also includes a refutation of evolution based on the second law of thermodynamics.

MORRIS, H. M. *The Scientific Case for Creation.* San Diego: Creation-Life Publishers, 1977.

A brief introduction to scientific creationism without references to the Bible. An excellent book to give colleagues or students for critical review.

WHITCOMB, J. C., and Morris, H. M. *The Genesis Flood.* Grand Rapids, Mich.: Baker Book House, 1961.

This book contains a documented analysis of geological aspects of the creation evolution question. It argues for a recent creation and a global deluge as the cause of sedimentary rocks and fossils. This book is considered the seminal work for scientific creationism.

[3]

Theories, Facts, and Gods: Philosophical Aspects of the Creation-Evolution Controversy

A. D A V I D   K L I N E

T O understand the basic issues in the creation-evolution debate, it is helpful to distinguish those arguments that are mainly empirical in nature from those that are philosophical or conceptual. On the empirical side, creationists have challenged a number of tenets of the evolutionists — for example, the thermodynamical possibility of the spontaneous origin of life or the accuracy and consistency of widely used techniques for dating geological strata. (Many of these issues are discussed in subsequent chapters of this book.) On the philosophical side, the areas of contention range beyond the theory of evolution to the nature of science itself. This chapter discusses four of these philosophical issues: (1) the relationship between theories and facts, (2) evidence for laws, (3) the relevance of "falsifiability" to evolutionary theory, and (4) the alleged place of God in scientific explanations. The discussion indicates why creationist arguments present an inconsistent and inaccurate view of the nature of scientific knowledge.

THEORIES OR FACTS

There is considerable and vitriolic disagreement over whether evolution is a theory or a fact. At Iowa State University, for example, such an interchange between a biology professor and his student made the campus newspaper. The student apparently pointed out with some delight that the theory of evolution was only a theory; the professor with some impatience insisted that it wasn't a theory but a fact. Speaking to an evangelical group in Dallas, Texas, President Ronald Reagan gave this opinion: "Well, it is a theory. It is a scientific theory only, and it has in recent years been challenged in the world of science — that is, not believed in the scientific community to be as infallible as it once was."[1]

The disagreement is not restricted to contrary students and professors or admitted anti-intellectuals. Among vocal "authorities," the issue has been kept alive by Stephen Jay Gould's recent paper "Evolution as Fact and Theory"[2] and Duane Gish's creationist response.[3]

At the quick of the issue is an equivocation on the meaning of the term *theory*. A tip-off to this is that the theory-fact distinction is alien to scientific discourse. The distinction comes from ordinary or prescientific discourse, in which facts refer to statements that are known for certain, or that are indubitable or obviously true, whereas theories refer to statements that are speculative and not well established. That Jack Ruby shot Lee Harvey Oswald is a fact. Many, actually millions, saw the shooting on television. The event engendered many theories as to Ruby's motivations. Was he part of a larger plot to silence Oswald? Was he simply a man intent on avenging John Kennedy's death? Or what?

Let us label the ordinary sense of *theory*, theory$_o$. Now, as has been pointed out, if something is theory$_o$, then it is not well established.

In the scientific context, however, the contrast is between theories and data. We know, for example, that human memory appears to have a practically limitless capacity for storing information. But how (by means of what mechanism) do we retrieve information from memory? Some especially elegant experiments by Saul Sternberg hint at the answer.[4] His work also provides a clear illustration of the theory-data distinction.

Sternberg gave his subjects a short list of digits to memorize — called the "positive set." Then he gave them an additional digit — the "test stimulus." The subject's task was to determine quickly and accurately whether the test stimulus was in the positive set or not.

Figure 3.1 is an idealized visual representation of Sternberg's data. The actual data are a set of results of the form: subject 1 had a reaction time of t on positive set p of size s, where t, p, and s are particular values.

Sternberg proposed a theory to explain the data. He suggested that, at least in simple cases like the described task, memory retrieval proceeds by a *serial* and *exhaustive* search. That is, the subject internally represents the set and the test stimulus, then compares the test stimulus against the first item in the positive set for a match. He then proceeds to the next item and so on until the test stimulus has been compared with every item in the positive set. The subject goes through the entire procedure whether a match is found or not.

Notice that the theory is not what we would normally expect — that is, that the search would be terminated as soon as a match occurred. Rather, the data

FIG. 3.1.   Graph of Sternberg's data.

support the theory that an *exhaustive* search occurs, since reaction times increase as the size of the positive set increases,[5] and since whether the test stimulus is actually in the positive set or not has no effect on the mean reaction time. Only the size of the positive set is relevant.

Several general points on the theory-data distinction are in order. First, data are typically used to support or justify a theory. Exactly how the data must be related to a theory in order to provide support is a complicated issue in inductive logic and, fortunately, one with which we need not be concerned here. Second, since it would be pointless to use *a* to justify *b* if *a* were less well established than *b*, typically the data are better established than the theories they support. Third, nevertheless, data are not certain or indubitable. The data are often discovered to be mistaken. The source of the error can be faulty instruments, false auxiliary assumptions, poor experimental design, and so on.[6] The point is that for data to be used in a justifying role requires that the data be correct, not that they be certain or indubitable.[7] If the data are correct and if they are related to the theory in the proper way, they provide some justification for the theory. Finally, justification is a historical as well as logical process. Although initially data are supposed to be much better established than the theory they support, they need not remain so. As the theory becomes supported by numerous data and explains more and more, our confidence in it rises. It can rise to such an extent that there is little difference between our confidence in the data and our confidence in the theory. Theories can come to be well established, indeed, very well established. Special relativity and transmission genetics are examples of such theories.

This last point is the important one for understanding the spat over whether evolution is a theory or fact. Let us label the scientific sense of *theory*, theory$_s$. The crucial conceptual truth is that to claim that a theory$_s$ is well established, even beyond a reasonable doubt, is not a contradiction.

Then how shall we speak of evolution? All should agree with the creationists that it is a theory, in that evolution is a systematically related set of regularities that allegedly explain numerous and diverse phenomena.[8] But the creationist concludes straightaway that evolution is not well established, a move made plausible only by an equivocation on the meaning of *theory*. The creationist's conclusion follows only on the supposition that evolution is a theory$_o$. But when biologists call evolution a theory, they mean that it is a theory$_s$. And on that reading the creationist conclusion does not follow. One cannot reason that since the theory of evolution is a *theory*, it is not well established.

The creationist may admit that the evolutionary account is a theory$_s$, yet not a well-established one. This point could be true, of course; but if so, it becomes an *empirical*, not a *philosophical*, point. The claim that evolution is not well established thus requires empirical justification.

REPEATABILITY OF EVENTS

Creationists charge that evolutionary theory has a special difficulty in conforming to one of the canons of scientific method. As Duane Gish notes: "Another criterion that must apply to a scientific theory is the ability to repeatedly observe the events, processes or properties used to support the theory. There were obviously no human witnesses to the origin of the universe, the origin of life, or in fact to the origin of a single living thing."[9] Gish's point, which is widely echoed in the popular creationist literature, seems to come down to this. Consider some scientific regularity—for example, that arsenic poisons or that copper conducts electricity. To have confidence in the truth of such a regularity or law, the creationists argue, one must be able to repeatedly observe instances of it. In the case of the regularities that make up the theory of evolution, it is in principle impossible to observe instances of them, since evolutionary history consists of a unique series of events.

This argument has two flaws.[10] First, the canons of scientific method do not require that one be able to observe the instances of the regularities or the particular event in order to justify one's belief in them. It is acceptable and standard practice to justify belief in a regularity by deducing certain observable consequences from it and certain auxiliary assumptions. The same procedure holds for establishing the occurrence of particular events. To insist that instances of the laws be observed would eliminate not only much of the theory of evolution but nearly the whole of modern physics. The kind of reasoning illustrated by the Sternberg experiments is standard fare in scientific inquiry. Sternberg did not observe the serial exhaustive search said to take place in the brain but observed consequences of it, namely, reaction times. Similarly, one does not observe the interaction of elementary particles but, rather, certain consequences or effects of those hypothesized interactions.

Second, even when instances of a regularity are observable—for example, Snell's Law ($\sin i / \sin r$ = constant)[11] or the gas law ($PV = NrT$)[12]—one cannot repeatedly observe the *same* events being instances of the regularity simply because every event is unique. No event occurs twice. So if we adhere rigorously to Gish's demand, no law can be confirmed.

Of course, given some notion of relevant similarity, relevantly similar events can occur an indefinitely large number of times. It is by observing such event *types* that scientists confirm observational laws. But there is no reason that the evolutionist cannot meet this demand. He claims that the same regularities hold now as in the past. Of course, he cannot observe past instances of the regularities. That is a trivial truth. Neither can he observe past instances of Newton's laws. But he can confirm Newton's laws in the laboratory and in nature. The same situation holds for the evolutionary regularities that have observable instances.[13]

In summary, Gish's supposed canon of method is far too strong. If confirmation requires the observation of instances of regularities, most of science will not meet the requirement. Even if we restrict the requirement to observational regularities, it is still too strict since events do not recur. If we weaken the requirement to event-types then there is no a priori reason that evolutionary regularities cannot meet the condition.

FALSIFIABILITY

The creationists' main criticism of evolutionary theory is the bold challenge that the theory is not scientific. This claim has had a powerful rhetorical effect among creationists. Scientists, they argue, have painted themselves as hard-nosed, no-nonsense fellows who claim that the weak-willed creationists have failed to properly limit religion and hence have tainted the truth. So the claim that evolutionary theory is a metaphysical system or, worse yet, an alternative religion is the ultimate criticism.

The creationist charge is based entirely on the work of the philosopher of science Karl Popper.[14] Over a half century ago Popper was impressed by the contrast between the "scientific" status of Marx's theory of history, Freud's psychoanalysis, and Alfred Adler's individual psychology on the one hand and Einstein's theory of gravitation on the other. He set for himself the problem of "demarcation," or the problem of defining when a theory should be regarded as scientific. Popper's answer is that the mark of a scientific theory is falsifiability.[15] In other words, theories are scientific if they prohibit some occurrences. There must be some observable events such that if they were to occur the theory would be false. Sternberg's theory, for example, could be falsified if subsequent data gave a different graph from that shown in Figure 3.1. As a matter of historical record, Popper believes that the theories of Marx, Freud, and Adler failed to meet the criterion, whereas Einstein's theory did. He also stated that "Darwinism" was not falsifiable and therefore not scientific.

The creationist argument against the scientific status of evolutionary theory can be reduced to a simple syllogism: *premise one,* falsifiability is the criterion for scientific status; *premise two,* the theory of evolution is not falsifiable; *conclusion,* the theory of evolution is not a scientific theory. To evaluate this argument, three points must be considered:

1. Creationists defend the premises by nothing more than an appeal to Popper's authority. Popper *said* that falsifiability is the mark of a scientific theory. Popper *said* that the theory of evolution is not falsifiable. Nowhere in the creationist literature do we find a defense of falsificationism or Popper's evaluation of evolutionary theory. We do not even find a clear statement of Popper's ideas. What Popper actually says about "evolutionary theory" is that "Darwinism is not a testable scientific theory, but a metaphysical *research programme* — a possible framework for testable scientific theories."[16] Nowhere do creationists demonstrate that what Popper means by "Darwinism" is what contemporary biologists mean by "the theory of evolution." Perhaps the current theory is one of those testable theories that fall within the Darwinian framework.

The creationists' crude appeal to Popper's authority can perhaps be understood, though not justified, when it is noticed that many anticreationists also accept Popper with no questions asked. Popperianism appears to be gospel among many scientists. Efforts by scientists to counter the nonscientific components of our culture, such as astrology and extrasensory perception, have typically been fought under the banner of falsificationism. Here, too, the banner has simply been borrowed, not examined.[17]

2. If one's arguments rest merely on appeals to authority and the authority happens to change his mind, then one is left, as they say, holding the bag. Unfortunately for creationists, that is precisely the present situation. Popper's

recent comments clearly indicate that he believes that the contemporary theory of evolution is testable.[18] Of course, he may be wrong. But what we need is an argument to that effect.

3. For the purposes of argument, let us suppose that falsifiability is the mark of a scientific theory. Present in this supposition are two embarrassing points that the creationists have overlooked. First, the bulk of the creationists' objections to evolutionary theory are straightforward empirical objections. For example, creationists claim that evolutionary theory is incompatible with our knowledge of thermodynamics and that evolutionary theory is incompatible with the fossil and sediment-layer records. It is by means of such objections that creationists attempt to establish their scientific expertise and to refute the theory of evolution.

Whether these critical claims are correct is not the present issue. (They are examined in subsequent chapters.) The present point is that if creationists claim that observable evidence actually refutes a theory, then they must think that it is falsifiable. Therefore, the creationists cannot claim to have given evidence that the theory of evolution *is false* and also that it *is not falsifiable*. Second, is creationism itself a scientific theory? Despite some disagreement among creationists on this issue, most would say that it is. The following remarks are addressed to those who believe creationism is a scientific theory and that falsifiability is the mark of such a theory.

The obvious question is whether creationism is falsifiable. I shall be suggesting that in a sense it is not. My remarks could be understood as encouraging creationists to state what those observable occurrences are, which, if they were to happen, would constitute counterevidence to their theory.

To determine whether creationism is falsifiable, let us consider first a possible theory called "originism," which is amazingly similar to creationism in that it denies almost every claim of evolutionary theory. According to originism, for example, the universe is about 10,000 years old, plants and animals appear in the universe as distinct kinds, the earth's history contains a massive catastrophic flood, and so on. Suppose that originism is just like creationism except that it is "naturalized." Everywhere that creationism talks about such and such being supernaturally created, originism talks about such and such appearing or occurring. Originism, then, is just like creationism except that it has been purged of nonnatural creative activities and, of course, creators.

It is obvious that originism is falsifiable. There are possible fossil records, or sedimentary records, or carbon-dating results that would refute originism. Since originism is "contained" in creationism, creationism will also, at least in principle, be falsifiable. That is, the same results as just stated could falsify creationism.

But given that originism is a simpler view than creationism, on standard methodological grounds it should be preferred unless creationism has additional observational consequences — consequences in addition to those of originism. I suspect there are none. For those who disagree, it will be instructive to have the observational consequences clearly stated. The next section formulates a dilemma for anyone who takes up this challenge.

GOD AND SCIENTIFIC EXPLANATION

Is it possible, in principle, for concepts like God, creator, and creative process to play a role in scientific explanations? As we have seen in the previous discussion, the answer depends on what you mean by "God," "creator," and "creative process." In particular, are these concepts understood in such a way that they play a role in a theory having observational consequences that the theory would not have without them?

This very general response is correct, but it makes it appear that the answer to the question is more open than it really is. It is not up to a specific speaker of a language to create for words whatever meanings he wants. Lewis Carroll put the point humorously in *Alice in Wonderland:* "That's a great deal to make one word mean," Alice said in a thoughtful tone. "When I make a word do a lot of work like that," said Humpty Dumpty, "I always pay it extra."

Consider the concept of God. Within the standard Judeo-Christian tradition, one that scientific creationists accept and want to be identified with, God is a supernatural being that is all good, all knowing, and all powerful. The essential question, therefore, is whether this specific God can play a role in scientific explanations. There appear to be logical reasons why it cannot.

God's nature, being supernatural, is unlike natural entities such as electrons, genes, apples, and societies. Entities in the natural order interact with one another through efficient causal relations. Very generally in efficient causal relations the entities are spatially and temporally contiguous and undergo a transfer of energy from one to the other—for example, a hammer striking a nail causes the nail to move into the wood. The heart of scientific explanation is providing the efficient causes of events or phenomena.

Now it should be clear that there is a severe logical tension in the claim that God could play a role in scientific explanations. The most plausible reason for allowing God as a scientific entity is that God enters into efficient causal relations. But if an entity enters into efficient causal relations, it can be understood as natural, not supernatural. God can acquire a scientific status only by abandoning his supernatural status. For those within the standard Judeo-Christian tradition, that price must be judged too high.

The philosophical arguments used by creationists to refute the theory of evolution—that it is merely a theory, that it is about nonrepeatable phenomena, that it is unfalsifiable—are woefully inadequate. The dust raised by these arguments was agitated by various confused or uncritically held views about the nature of science. With the dust removed, we can turn to the empirical issues discussed in the next six chapters.

FOR FURTHER READING

HEMPEL, CARL. *Philosophy of Natural Science.* Englewood Cliffs, N.J.: Prentice-Hall, 1966.
    A brief, lucid discussion of some of the central topics in the philosophy of science.

HULL, DAVID. *Philosophy of Biological Science*. Englewood Cliffs, N.J.: Prentice-Hall, 1974.
Surveys philosophical problems in the biological sciences.
KLEMKE, ELMER; HOLLINGER, ROBERT; and KLINE, A. DAVID, eds. *Introductiory Readings in the Philosophy of Science*. Buffalo: Prometheus Books, 1980.
Contains essays from contrasting viewpoints on standard topics in the philosophy of science.
POPPER, KARL. *Conjectures and Refutations*. New York: Harper & Row, 1963.
Essays on topics in the philosophy of science, including the demarcation problem.
SCHILPP, PAUL ARTHUR, ed. *The Philosophy of Karl Popper*. Books I and II. La Salle, Ill.: Open Court, 1974.
Essays on Popper with an important autobiography in which Popper discusses evolution.
SUPPE, FREDERICK, ed. *The Structure of Scientific Theories*. Urbana: University of Illinois Press, 1974.
A collection of essays on the nature of theories along with a critical introduction by the author.

II

The Realm of Science

[4]

Astronomical Creation: The Evolution of Stars and Planets

JANE L. RUSSELL

ASTRONOMY provides a clear example of the contrast between scientist and creationist. From astronomical observations and well-established scientific theories modern astronomers have concluded that the universe has evolved in definite ways during a period of several billion years. Relying ultimately on Genesis, creationists propose that the universe is only a few thousand years old and came into existence essentially in its present form.

In 1972, as director of the Institute for Creation Research, Henry M. Morris published *The Remarkable Birth of Planet Earth,* a book that discusses astronomical evolution as well as other issues in the creation-evolution controversy. Setting the stage for his discussion, he explains, "I personally have become thoroughly convinced that the Biblical record, accepted in its natural and literal sense, gives the only scientific and satisfying account of the origin of all things."[1] In accordance with this viewpoint, Morris's discussion "is primarily approached from the Biblical point of view, and assumes throughout that the Bible is the Word of God, divinely inspired and, therefore, completely reliable and authoritative on every subject with which it deals."[2]

Morris's detailed discussion of astronomy in Chapter 6, "The Puzzling Role of the Stars Above," begins with the disagreement between astronomers and creationists. According to Morris, astronomers' theories of the evolution of the universe seem to constitute "a tongue-in-cheek charade."[3] Because the universe has appeared the same throughout the long history of observational astronomy, Morris argues that observationally minded astronomers ought to accept that the universe was specially created in its present form: "The problem is completely settled, of course, by the Scriptures."[4]

Morris devotes the rest of the chapter to explicating biblical passages and to considering the possible purposes of stars. He notes, for example, the apparent biblical contradiction that although God created light on the first day of creation, he did not create the sources of light (the sun and stars) until the fourth. Morris concludes, "There is no way now of determining the nature of this initial source of visible light, since it was later delegated, as it were, to the heavenly bodies. It may well have emanated from the theophanic presence of God Himself."[5]

One modern astronomical finding that Morris does accept is that stars are millions or billions of light-years away from the earth. Since it takes light one year to travel the distance of one light-year, astronomers conclude that light from these stars must travel for millions or billions of years to reach the earth. Morris admits this creates a problem, however, for "if the stars were made on the fourth day, and if the days of creation were literal days, then the stars must be only several thousand years old."[6] He resolves the problem by concluding that God created not only the stars but also all the light *between* the stars and the earth. It is even possible, according to Morris, that this was the light created on the first day. That is, God may have created starlight three days *before* he created the stars themselves.[7] In discussing the possible purposes of stars, Morris presents some of his more unusual ideas:

> There are a number of Biblical references indicating that in some way the stars may actually participate in human battles (Numbers 24:17; Judges 5:20; Revelation 6:13; 8:10; etc.). Such passages may all be simply figurative, but then again they may not. . . . The long fascination of men of all nations with pagan astrology can only be understood if it is recognized that there is some substratum of truth in the otherwise strange notion that objects billions of miles away could have any influence on earthly events. Certainly the physical stars as such can have no effect on the earth, but the evil spirits connected with them are not so limited.[8]

Morris's book is a prime example of the creationists' avowed desire to reinterpret science from a biblical viewpoint. As he says, "real creation necessarily involves creation of 'apparent age.' "[9] The universe may look old to scientists, but the Bible tells us it is not.

Astronomers, on the other hand, have spent great effort trying to remove astronomy from an occult area of study to a scientific one. They have been hampered by the size of their subject — that is, the entire universe — and by the difficulty of obtaining observations about many areas of it. After all, astronomy is an observational rather than a laboratory science. There are no controlled experiments in astronomy; astronomers cannot put stars in a laboratory and "poke" them to observe what happens. With the exception of satellite observations, they are confined to observations made from the home planet through what is at best a capricious atmosphere. Nevertheless, astronomers have made enormous progress in understanding the activities of stars and galaxies, especially within this century. Fitting modern observations into a network of long- and well-established scientific principles, twentieth-century astronomers are providing a clearer and clearer account of how the universe developed into its present state.

Astronomers' overall conception of the universe rests largely on a number of observations not mentioned in Morris's book. The science of spectroscopy, developed in the nineteenth century, has been exceedingly useful in the twentieth. Spectroscopy has shown that each element, when incandescent, gives off its own characteristic "spectrum" of colors. Hence, by analyzing light from a star, astronomers can determine the star's chemical composition. In this way helium was discovered in the sun before it was discovered on earth. As we shall see, the chemical composition of stars is an important aspect of our understanding of stellar evolution. Moreover, by studying the "shift" in the colors of light coming from a star, astronomers can determine the speed at which the star is moving toward or away from us. The effect is similar to the change in pitch of a train's whistle, which is higher when the train is approaching a listener and lower when the train is receding. In the case of light, it is the color that changes, and the greater the speed, the greater the change or shift in color. The result of such observations is that all observable stars are moving rapidly away from one another. The current speeds of stars and the distances between them provide the basis for the further conclusion that the various parts of the universe were thrown out by an enormous explosion that must have occurred some thirteen to twenty billion years ago. Indeed, this is the main idea of the so-called big bang theory.

Although astronomers have little or no evidence of what may have happened or existed *before* the big bang, they have ample evidence of what has happened between then and now. To begin with, conditions within that long-ago explosion would have been vastly different from the current state of the universe, and astronomers therefore conclude that enormous changes have taken place over an immense period of time. By correlating specific observations with well-established scientific theories, astronomers have been able to formulate increasingly detailed theories of the nature of those changes. One set of observations, for example, is that there are many different kinds of stars, and we shall see how such observations bear on the theory of stellar evolution. The two natural laws central to astronomers' explanations of the development of the universe are the law of universal gravitation and the perfect gas law. The lifetimes and evolution of stars are said to be the result of a continuing interaction between the respective activities described by these two laws.

The law of universal gravitation describes the mutual attraction that exists between any one object and every other object in the universe. The magnitude of that force increases with the mass of the objects considered (that is, objects twice as massive will have a force of attraction twice as great), and the force decreases with the square of the distance (that is, if the distance between the objects is made three times greater, the force decreases by three times three and so would be only one-ninth as great). Objects on the surface of the earth are dominated by the gravitational attraction of the earth because the earth is very massive and very close. Although the earth's attraction is the gravitational force most obvious to us, gravity also acts between the objects themselves—people are attracted by furniture; buildings are attracted to each other; the tennis player is attracted to his racquet, the tennis ball, the net, and his opponent. In all these examples the force obviously is miniscule compared with the force between each object and the earth. But consider the attraction between bodies of nearly the same mass that are a great distance from any more massive object. This is the usual situation in astronomy. A local example is the gravitational attraction that holds the earth and moon in orbit around each other. As each tries to move off in a straight line, their mutual attraction pulls them back so that they move in nearly circular motion. Gravity binds double stars so that they continue to orbit one another, and gravity holds together clusters of galaxies. Although the weakest of the known natural forces, gravity holds the universe together.

The ideal gas law also has great bearing on events in the universe because most of the matter is in a gaseous state and behaves as a so-called ideal gas. That means that the relationship of the temperature, pressure, and density of this matter is predictable. Increasing the density of a gas increases the pressure it exerts. When a balloon is squeezed, for example, the density of the gas in it is increased, and the resultant increased pressure produces bulges in the other areas of the balloon. An increase in the temperature of an ideal gas correlates with an increase in the pressure. Thus, if a sealed container of gas is heated and the interior gas temperature increased, the container will become distorted or may even burst because of the increased gas pressure.

The stage is now set to explain the formation of stars. Gravitational forces of attraction between gas particles will tend to collapse any conglomeration of gas. But any such pulling in of a gas cloud increases the density and thus the pressure of the gas. This increased pressure would in turn tend to disperse the cloud. When these forces find a balance point at which the gravitational force inward is just equal to the outward gas pressure, the gas cloud is said to be in "hydrostatic equilibrium." This is the condition of the gases in the sun, and observations confirm that the sun is not changing its size significantly. The gas law can be used to calculate the temperature at any point in the sun. One first calculates how much gas pressure is coming from the portion of the sun interior to that point to balance the gravitational force (weight) of the portion exterior to that point. That gas pressure and density enable us to calculate the temperature.

With this information at hand, let us consider stellar evolution. The first step is to find enough matter to form a star, and possibly planets. There are scattered throughout our galaxy, and observed in other nearby galaxies, giant clouds of gas, primarily hydrogen with some helium and a few other elements mixed in, often in the form of dust grains. By earth standards, the gas in these giant clouds is more tenuous than the best vacuums achievable in a laboratory. If, for example, we were to take the amount of the earth's atmosphere that can be held within two hands—say, about the size of a baseball—and expand it until it had the same density as one of the interstellar clouds, that handful of air would fill a box the size of Ohio and one mile deep. But these clouds, known as nebulae, are so large that they may contain up to several hundred times the mass of our sun; larger ones may contain several thousand times that mass.

Every particle in such a cloud has its own motion, evident as low-level turbulence. If the turbulence combines so that a small compression is caused in one area of the cloud, it may be sufficient to give gravity the advantage over gas pressure and the cloud will begin to collapse. The same type of collapse can also be started by an outside influence such as a shock wave (like the compression wave from an explosion) passing through the cloud. In any case, once the collapse begins, the cloud's behavior is predictable as long as the known natural laws hold.

With increasing density from the collapse, the cloud will begin to heat up. Some of the heat will be dissipated as infrared radiation; the rest will go into heating the cloud, at this stage called a "protostar." The pressure from the heated gas will still not be sufficient to halt the collapse, and it will continue. With the aid of infrared techniques developed in the last decade, astronomers have been able to observe some objects that appear to be protostars shrouded in gas and dust. When the interior temperature reaches about 3 million °K (about 5.6 million °F), the star "turns on." Up to that point all the radiation from the collapsing cloud is due to the heating from gravitational collapse. When the density and temperature at the core of the protostar become high enough, however, nuclear fusion will begin and will continue for most of the life of the star.

Nuclear fusion is the process whereby light atomic nuclei, such as hydrogen, combine to form heavier nuclei, with energy being released in the process. Fusion produces energy as long as nuclei heavier than iron are not involved. The fusion of iron requires energy to be put into the system, not removed. Fusion is the process behind the hydrogen bomb. It should not be confused with fission, the process central to the operation of the atomic bomb, which produces energy by breaking up heavy nuclei like uranium into lighter nuclei. The fission process produces energy as long as the nuclei involved are heavier than iron.

Nuclear fusion in the sun and other stars uses the lightest element, hydrogen, which makes up about 75 percent of the mass of the stars, and produces helium, the second lightest element. This reaction takes place in several ways, different ones operating most efficiently at different temperatures and using different catalysts. They all produce the same result—energy. The energy comes from the mass lost in the transition, 0.007 of the mass of the original hydrogen being lost in the transformation to helium. This missing mass has been converted to energy according to Einstein's famous equation ($E = mc^2$).

Once fusion begins in the stellar interior, the gas pressure and gravitational forces finally reach a balance, and for a long time the star is a remarkably stable system. If for some reason the interior temperature were to drop slightly, the rate of conversion of hydrogen to helium would decrease, but the resulting lower pressure would allow a little gravitational contraction, which would increase the central density and temperature and thereby immediately restore the energy production rate. Forces of gravity and gas pressure thus act as a thermostat to maintain energy production. This stable period, when hydrogen is being made into helium, is the "main sequence" lifetime of a star and encompasses about 90 percent of the time that a star is a recognizable entity. In general, the more massive a star, the hotter it is. Even though it has more fuel available than less massive stars, because of its higher temperature a higher mass star uses its fuel at a much higher rate, so that the massive stars have the shortest lifetimes. The largest stable stars are about fifty times the mass of our sun and use their fuel so rapidly that they cannot live longer than a few million years. Our sun's lifetime will be about ten billion years. The smallest mass stars have about one-tenth the mass of our sun and use their fuel so slowly that they will last hundreds of billions of years.

When the hydrogen fuel in the core of a star is exhausted, the star begins to change. The smallest mass stars will most likely swell, cool off, collapse, and fade away. The stars of about the same mass as our sun will go through what is called a "red giant" phase before dying out. In this phase of its evolution, the interior of the sun will begin to convert helium to carbon, another fusion reaction that will heat the interior to about 100 million °K. The outer layers of the star will swell enormously and cool at this time. As a red giant, the sun should swell to just about fill the earth's orbit; the earth, or what is left of it, will orbit in the outer layers of the sun. Most likely the sun will lose about 10 percent of its mass from the upper layers and then begin to collapse in on itself. The collapse will not be halted at the sun's main sequence size, because the internal pressures of fusion reactions will no longer be a factor. Hence, the sun will ultimately shrink to about the same size as the earth. Because its much larger mass will be concentrated in such a small volume, the sun in this final stage, known as a "white dwarf," will have incredibly high density, so high that just a teaspoon of it would weigh approximately five tons on the earth today.

The most massive stars will become red giants but will not then shrink peacefully to a smaller size. They will "supernova" and literally blow themselves apart, returning most of their mass to the interstellar clouds. What mass is left will be compressed into neutron stars or in some cases into black holes, some of the more exotic objects studied by astronomers. The matter that is returned to the interstellar clouds is of great interest to stellar evolution because it is then available for the formation of a new generation of stars. It enriches the interstellar medium among the stars with heavier elements formed during the lifetime of the star and during the supernova explosion itself, and this enrichment affects the composition of stars formed later. Moreover, the supernova produces a powerful shock wave, or echo, in the interstellar clouds, which tends to trigger the condensation of new stars. Thus stellar evolution comes full circle, from gas cloud to gas cloud, so to speak.

This theory is based on the assumption that the natural laws that we know are true on the earth are also valid throughout the universe. These laws can be used, with the aid of high-speed computers, to calculate the state of the star for a series of different epochs, thus enabling astronomers to simulate stellar evolution. The results of these calculations can then be compared with observations. It has been possible in this way to obtain good agreement between theory and observation, and this is taken as evidence that our theoretical models of stars, though somewhat simplified, represent real stars and their behavior. Such models predict the birth of new stars in gas and dust clouds. Because of the length of the time scales involved, one would not expect to observe a large number of new stars appearing across the sky. But there should be some new ones in reach of modern telescopes, and at least one stellar birth has apparently been recorded. In one dust cloud, high rates of infrared emission were observed. Infrared radiation is considered a sign of a collapsing cloud obscured by dust and heating up until it has sufficient temperature and pressure to begin nuclear fusion and become a main-sequence star. Comparing photographs taken at different times within this century, one sees a new star shining in the later picture but not in the earlier one. The early picture should have shown the star if it had been shining then. The star has continued to shine since, so it cannot be mistaken for a nova, which would fade after a few days or weeks. Hence, although astronomers continue to monitor and test various explanations of the photographs, the presence of the star may eventually provide direct and immediate support for the reality of the processes of stellar evolution.

The chemical composition of the very massive stars also provides evidence for the theory. The most massive stars, which have the shortest lifetimes and must therefore be young objects, show a higher proportion of heavy elements in their composition than older stars. This is predicted by stellar-evolution theory, since the younger stars would be expected to contain more matter that had been produced and then released by previous generations of stars.

Star clusters provide additional support to the theory. One would expect stars close enough together to be gravitationally bound into a star cluster to have been formed at about the same time from a very large, fragmenting cloud of interstellar material. Therefore, all the stars within the cluster should be about the same age, an expectation supported by the observation that they have the same chemical composition. Since the *age* of these stars is the same, one should expect to observe a direct correlation between their respective *masses* and stages of evolution. According to the theory, as we have seen, the characteristics of a star at any given time depend on how old and how massive it is. Hence, in the case of star clusters, where nature has in effect provided us with a group of stars of the same age, their different evolutionary stages should correspond to their different masses. Indeed, stars in such clusters *are* observed to have an orderly progression of evolutionary states such that the massive stars are further along than the less massive stars, and in many clusters the very low mass stars are still collapsing to the main-sequence state.

These are a few examples of how observations support the astronomers' theory of stellar evolution. For a more complete picture, the reader is referred to the list for further reading given at the end of this chapter.

To understand the formation of planets, we first need to discuss multiple star systems. These are double stars (two stars that orbit each other), triple stars, and even quadruple, quintuple, and sextuple systems. Of the sixty-seven known stars nearest the sun, which are our closest neighbors and most closely studied, thirty-two, or nearly 50 percent, are in multiple star systems. Of the remaining thirty-five stars, seven are suspected of having companions, which would raise the percentage of stars in multiple systems to over 60 percent. As observational techniques improve, more companions are found. The implication is that if stars can form in clusters and in multiple systems, they should also be able to form with planetary companions or with both stellar and planetary companions. Computer simulations, for example, have shown that in double stars it is possible also to have planets in stable orbits. And we know that our sun has an entourage of planets and miscellaneous bodies that orbit with it. Our sun is not extraordinary in any other way, so it seems reasonable to assume that at least some other stars also formed with planetary systems.

Direct evidence to support the existence of planets around other stars is lacking. Astronomy has just gained the necessary technology to search for planets outside our solar system within the 1970s, and the search is conducted on the edge of observational accuracy. But to assume that among all the known stars only our sun has planets is the type of exclusive statement astronomers have often made and often regretted. That the earth was the center of the solar system, that the sun was the center of our Milky Way Galaxy, or that the Milky Way was at the center of the local group of galaxies were all assumptions of exclusive places in the universe for our local system, and all have been proved wrong.

Planets orbiting other stars could not be observed directly; they would be lost in the glare of radiation from their suns. Thus the search must be conducted indirectly, for example, by carefully studying the motion of a star, which will be perturbed if it has an accompaniment of planets, or by monitoring the light from a star for variations in its brightness or velocity, which will also indicate the presence of planets. Several searches using different methods are now under way.

And how could planets form with a central star? The exact scenario of the formation of the solar system is still unknown. Astronomers have deduced a general idea of the events that is probably correct in its overall view but will undoubtedly need revision in its details as more data become available.

Let us consider again the collapsing gas cloud that formed our sun. As such a cloud began to collapse, it would have some rotation because of the motions of the particles that composed it. As the collapse proceeded, the cloud would rotate faster to conserve the angular momentum. As it collapsed still further and rotated still faster, the spherical cloud would begin to flatten into a disk, leaving a spinning disk of gas behind while the central cloud continued to collapse to the sun. Within the gas disk around the protosun there would be collisions among the particles. If their relative velocities at the time of collision were not too high, the particles would begin to accrete. And different concentrations of elements would develop at different parts of the cloud. The heavier particles of silicas and iron would continue in their orbits throughout. The lighter elements and ices would not be able to stay near the protosun and would remain in the outer, cooler regions of the gas disk. Thus the matter accreting nearer the sun would be mostly the rocky metallic elements such as those that make up the "terrestrial" planets nearest our sun. The outer accretions would maintain the original composition of the solar nebula, mostly hydrogen and helium and also some of the heavier elements such as those that make up the "Jovian" planets farther from the sun.

The continuing buildup of the mass of a few bodies surrounding the protosun would allow them eventually to sweep up much of the matter in their portion of the forming solar system. Planets would be cleaning up their own ring-shaped territory around the sun. When the sun finally began fusion reactions in its core and became a star, there would have been a brief time when it was brighter than it would be most of its lifetime. During that brief period, the sun would have emitted a large amount of radiation plus a significant solar wind (that is, small particles given off by the sun). Together these would sweep out the solar system, blowing away most of the debris to leave the sun and the planets remaining.

The moons orbiting the planets could have originated either in the breaking up of the accreting material into separate bodies in orbits around one another, or in the capture of material when the gravitational attraction of a planet pulled in a passing chunk of matter moving too slowly to escape. The moons most likely to be captured were those with the most eccentric orbits and those that moved "retrograde." (Retrograde refers to bodies moving east to west in their orbit or rotation as opposed to the west-to-east or "direct" motion that is usually found in the solar system.)

This general description is probably on the right track, but the details of a stellar formation that would put all the observed planets in their proper places have yet to be worked out. Astronomically, planets are minor details in the overall scheme of our galaxy and the whole universe, but studies of stellar evolution are beginning to deal with such details. Recent computer simulations of thousands of bits of debris orbiting a collapsing protosun have shown the formation of just a few planets around the sun to be feasible.[10]

This discussion has been at times superficial and lacking in detail, but when we consider that at the turn of this century astronomers did not even know what caused the sun to shine, we recognize that astronomy has come a long way in explaining the characteristics and formation of stars. In continuing to study the universe, scientists will no doubt make errors and wrong assumptions before continuing along the correct path of knowledge. Nonetheless progress will be made. If future observations should contradict current theory, then current theory will be changed. That is the way of science, as M. Schwarzschild, astrophysicist at Princeton University, has noted:

> If simple perfect laws uniquely rule the universe, should not pure thought be capable of uncovering this perfect set of laws without having to lean on the crutches of tediously assembled observations? True, the laws to be discovered may be perfect, but the human brain is not. Left on its own, it is prone to stray, as many past examples sadly prove. In fact, we have missed few chances to err until new data freshly gleaned from nature set us right again for the next steps. Thus pillars rather than crutches are the observations on which we base our theories; and for the theory of stellar evolution these pillars must be there before we can get far on the right track.[11]

For Further Reading

ABELL, GEORGE O. *Exploration of the Universe.* New York: Holt, Rinehart & Winston, 1975.

Astronomy. Milwaukee, Wis.: Astromedia Corp., 1973 continuing.

BERMAN, LOUIS, AND EVANS, T. C. *Exploring the Cosmos.* Boston: Little, Brown, 1980.

KAUFMANN, WILLIAM J. III. *Galaxies and Quasars.* San Francisco: W. H. Freeman, 1979.

SHIPMAN, HARRY L. *Black Holes, Quasars, and the Universe.* Boston: Houghton Mifflin, 1980.

Sky and Telescope. Cambridge, Mass.: Sky Publishing Corp., 1941 continuing.

[5]

Interpreting Earth History

BRIAN F. GLENISTER AND BRIAN J. WITZKE

THE current activities of the small but vocal group of creationists is of great concern to scientists and educators particularly because of the possibility that the dedication of creationists, coupled with public apathy, might result in passage of legislation that would seriously impair our educational system and adversely affect support for internationally vital scientific research. Most creationists, like their scientific opposites, are striving to understand the earth as they perceive it. This chapter is written in the belief that evidence preserved in the earth's crust bears importantly on the debate between the two groups. Consequently, we have attempted to review creationist literature and to test as objectively as possible how available geologic data fit the creationist and scientific models for earth history.

To introduce the scientific interpretation of the geologic record, we present first our own views together with the virtually unanimous consensus of professional geologists. To introduce creationist interpretations of geology, we present an extensive survey of their literature. These two conflicting interpretations are then contrasted in summary form. Next, we evaluate some of the geologic data that seem to offer the most reliable bases for the interpetations of the physical and biological histories of the earth. Of necessity, this part of the discussion is somewhat technical in nature, but it includes citations of both classic and current geologic interpretations and thus facilitates access to relevant literature. We conclude with some observations about the differences between the conflicting models.

The contrast between creationism and science is both strong and significant. Creationism rests on a single model, the one revealed by study of the Bible. Thus there is little incentive to generate new information from either field or laboratory study, as reflected in the lack of legitimate geologic research by creationists. Instead, they use the geologic literature as a source of data to support their model of earth history. To scientists, many creationists appear to mislead by divorcing such geologic data from their context and by using them in an unfairly selective way.

This chapter has benefited from critical review by the editors and by colleagues J. A. Arnold, R. V. Bovbjerg, P. H. Heckel, G. Klapper, W. W. Nassichuck, S. L. Weinberg, and E. E. Zylstra.

In contrast, geologists, like scientists in general, are committed to the principle of *multiple working hypotheses*.[1] Rather than running the risk of becoming too attached to a single hypothesis or model, geologists prefer to consider several alternatives concurrently. Field and laboratory data are then sought to support or refute the competing interpretations. The assumption that continents have always occupied the same relative geographic positions, for example, was abandoned recently in response to overwhelming evidence for the alternative models of continental drift and seafloor spreading. However, certain other hypotheses, such as evolution, have been tested and supported repeatedly so that their rejection may now be accepted as statistically improbable. Science, then, is a dynamic, multiple-hypothesis discipline in which old hypotheses are abandoned in the light of new evidence, whereas creationism represents a static, one-hypothesis approach to earth history.

Because creationist literature is now so voluminous that point-by-point refutation is impractical, we review only those data and interpretations considered most significant in deciphering earth history. In doing so, we hope to demonstrate both scientific methodology and the basic impartiality of our discipline; we also assert that, despite some changing interpretations, most of the basic tenets of geology have remained inviolate for over a century.

SCIENTIFIC INTERPRETATION OF THE GEOLOGIC RECORD

Science has proposed, on the basis of simple methodology, that billions of years of earth history predate written records. By observing present-day geologic processes such as erosion or accumulation of sediment and their resultant products, and assuming only the invariance of the physicochemical laws of nature, scientists believe they can understand the manner in which ancient geologic features were formed. This uniformitarian principle will be considered in more detail in a later section, as it is pivotal to geologic as well as other scientific endeavor.

Most of the details of the physical and biological history of the earth are recorded in the layered (stratified) *sedimentary rocks* that compose the consolidated bedrock beneath 75 percent of today's land area and virtually all the oceans. Study of the distribution of these layers in space and time is the geological discipline of stratigraphy. Individual layers are formed by accumulation of particles derived by erosion of preexisting rocks, accumulation of the hard parts or organic material from animals or plants that lived at the time that the sediment was forming, and chemical precipitation from aqueous solutions. Transformation of the original soft sediment and enclosed organisms into a solid sedimentary rock proceeds by compaction and desiccation and by cementation of adjacent grains. These are well-understood processes readily observable today. Although sedimentary rocks may accumulate in water or on land, the greatest volume has been preserved in the oceans because sea level is the ultimate base of erosion; above that level, erosion is the dominant geological process, whereas beneath it accumulation of sediment prevails. Both the sediment that accumulates at any given place and time and the organisms present and available to contribute their hard parts and organic material to it are related to the environment. For example, the sediment and organisms that sink to form a layer on the floor of an alpine lake will be different from those accumulating on a shallow tropical marine shelf studded with coral reefs. The sedimentary rocks and their contained fossils thus attest to the climate and other environmental conditions that prevailed at the time the rock layer formed.

For over three centuries, since the publications of Nicholas Steno, geologists have recognized that successive layers of sedimentary rocks could not have formed simultaneously, but must have accumulated particle by particle and layer by layer. According to this principle of *superposition,* each bed is older than any overlying unit unless dislocated by structural movements. Sedimentary beds thus yield evidence of conditions in successive intervals of geologic time and provide a record of earth history.

In many parts of the earth, sedimentary rocks can be observed in natural exposures such as the mile-thick succession in the Grand Canyon of the Colorado River. Tens of thousands of wells are drilled annually in the search for water or hydrocarbons, some to depths in excess of 30,000 feet; each provides information on the strata penetrated. Any dislocation or deformation of the originally horizontal sedimentary layers is generally obvious in outcrops (surface rock exposures), and the contours of even deeply buried strata can be traced and structural complications recognized through a number of geophysical techniques. Outcrops, wells, and geophysical data thus reveal innumerable *stratigraphic successions* (that is, sequences of sedimentary rocks that contain fossils) in which chronologic relationships can be determined objectively.

No single sequence of sedimentary rocks represents a complete record of earth history, and local gaps in the record (*unconformities*) are common. Even so, geographically separate rock sequences can be correlated in time by several independent methods (for example, stage of organic evolution, radiometry, tracing of volcanic ash beds, geomagnetism). Consequently, chapters of earth history that are unrecorded in one section can be inserted in their proper sequence on the basis of data derived from other sections. In this manner, geologists have reconstructed an impressively detailed composite history of the physical and biological evolution of the earth. The process is analogous to reconstructing a novel from numerous mutilated copies. Missing pages are the unconformities, and charred, crumpled, and torn pages have their analogs in sediments that have been altered, deformed, and broken by earth movements. The composite narrative is clear in both cases.

The great *antiquity of the earth* has become progressively clearer, especially since James Hutton concluded two centuries ago that most geological processes act slowly (by human standards) but that given sufficient time they can produce the preserved geological record. To test this idea, Hutton's contemporaries investigated such diverse phenomena as rates of accumulation of recent sediment and rates of introduction of salts into the oceans. Projecting these rates to account for the known thicknesses of sedimentary rocks and the salinity of the present oceans, they estimated the age of the oceans to be in excess of one hundred million years (at a time when most of the western world accepted 4004 B.C. as the date of creation.)[2] The great antiquity of the earth has been further verified with the discovery of the radioactive clock. Modern methods of radiometric dating are independent of older geologic dating techniques and provide a method for determining the age of most geologic materials and related events in earth history in absolute terms (that is, years before the present). These methods indicate that the age of the earth is approximately five billion years,[3] and that the oldest sedimentary rocks go back over three billion years (Figure 5.1).

Thus sequences of sedimentary rocks provide a composite physical record of three billion years of earth history. Traces or remains of the animals and plants (for example, tracks, shells, bones, wood) that lived when individual layers were accumulating are preserved in these sedimentary rocks as *fossils*. The process of fossilization can be observed in almost any area where sedimentary rocks are accumulating today. Mineralized structures such as shells and bones are most readily preserved, but soft plant and animal tissues may be portrayed in faithful detail as films of carbon. Even when soft tissues are not preserved, their general nature can be deduced from such features as muscle scars on shells or bones or from canals in bone through which nerves or blood vessels pass.

The record of life is full and clear for the last 600 million years of earth history, which make up the Phanerozoic interval (see Figure 5.1). For the preceding four billion years, the Cryptozoic interval, the record is somewhat vague, partly because of the microscopic size of most of the organisms alive at that time, but mainly because of the total absence of mineralized hard parts in organisms prior to the Cambrian Period, the beginning of the Phanerozoic. Development of these mineralized supports during the Cambrian not only permitted living things to achieve greater size than before and to develop new types of biological organization but also greatly enhanced the chances for fossilization in sedimentary rocks.

The fossil record affords an opportunity to choose between evolutionary and creationist models for the origin of the earth and its life forms. If the evolutionary model is valid, sequences of fossils through time will display "change with cumulative modification" (that is, evolution). For contrast, the creationist model explains the entire fossil record as a product of a single short-term global catastrophe, the biblical Flood; hence, no cumulative change "in the higher and intermediate categories of animals and plants" should be observed in fossil-bearing sequences, and no new groups may appear after the "creation." The record is clear. Groups of organisms appear successively in the record, and almost every biological family displays cumulative change when its fossil sequences are traced for significant intervals of geological time. This evolution is striking in most groups, for example, in the ammonoid cephalopods, which are discussed later in the chapter. However, in groups such as snails it may not be as clear, largely because the fossilizable parts of some organisms are generalized structures that cannot reflect complex changes in the soft tissues. The record is less satisfactory for organisms that were not easily preservable, especially for those that lacked hard parts or lived above sea level where erosion rather than sedimentation was the dominant sedimentary process. As a result the geologic record of shelled marine invertebrates reveals the most information.

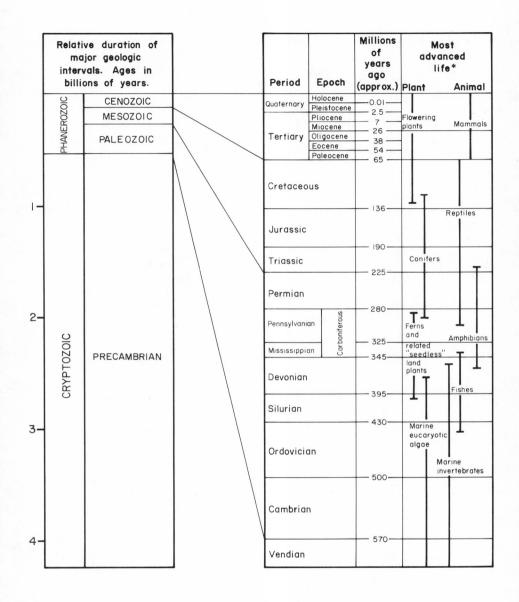

FIG. 5.1. Time scale of earth history. Vertical lines in the right column show only the intervals of dominant importance, not total ranges. All groups shown persist to the present day. From D. L. Eicher and A. Lee McAlester, *History of the Earth* © 1980. Adapted by permission of Prentice-Hall, Inc., Englewood Cliffs, N.J.

In addition to illustrating the progressive modification within individual groups of animals and plants, the geologic record reflects both the origination of new groups and the extinction of others. Figure 5.2 shows the distribution in time of some well-known groups.

CREATIONIST INTERPRETATION OF GEOLOGY

Creationism encompasses far more than the simple rejection of organic evolution. It advocates a drastic reinterpretation of virtually all aspects of the geologic sciences, particularly those that relate to earth history, which creationists believe is beyond the scope of science. Paradoxically, much of "scientific" creationist literature is devoted to earth history. Creationists largely reject the scientific methodology used in geologic studies and instead depend on revelation and biblical interpretation, as summarized by the leading creationists J. C. Whitcomb, Jr., and H. M. Morris:

> Any real knowledge of origins or of earth history antecedent to human historical records can only be obtained through divine revelation. . . . It is manifestly impossible ever to really prove, by the scientific method, any hypotheses relating to pre-human history. . . . [instead creationists seek] to build a true science of earth history on the framework revealed in the Bible, rather than on uniformitarian and evolutionary assumptions . . . [by] letting the Bible speak for itself and then trying to understand the geological data in light of its teachings.[4]

Morris, who is director of the Institute for Creation Research, has further emphasized that " 'Science' can only deal legitimately with *present* processes . . . we must depend *completely* on divine revelation."[5]

Although creationists claim that "creationism can be studied and taught in any of three basic forms," namely, "(1) scientific creationism, (2) Biblical creationism, and (3) scientific Biblical creationism,"[6] the geologic interpretations inherent in these forms have their basis in a literal interpretation of Genesis. Thus the outline of earth history presented by scientific creationists is basically the same as that presented in the biblical creationist model. The differences are primarily semantic. It is not surprising, then, that creationists claim that "all the statements in the Bible are consistent with scientific creationism."[7] Scientific creationists assert, for example, that virtually all sedimentary rocks and their contained fossils were formed catastrophically in a "recent global hydraulic cataclysm."[8] Biblical creationists identify this cataclysm as the Noachian Flood recorded in Genesis. Geologic evidence for an ancient earth is also rejected by the creationists, who argue that "Biblical chronology" indicates the earth has existed for only a "very short time span":[9] "the earth is really only several thousand years old, as the Bible teaches."[10]

The creationist view of geology can be found in publications written by members of the Creation Research Society (Ann Arbor) and the staff of the Institute of Creation Research (San Diego). Some of the key points that recur in the creationist literature are summarized below:

1. Earth history has been dominated by catastrophism, and uniformi-

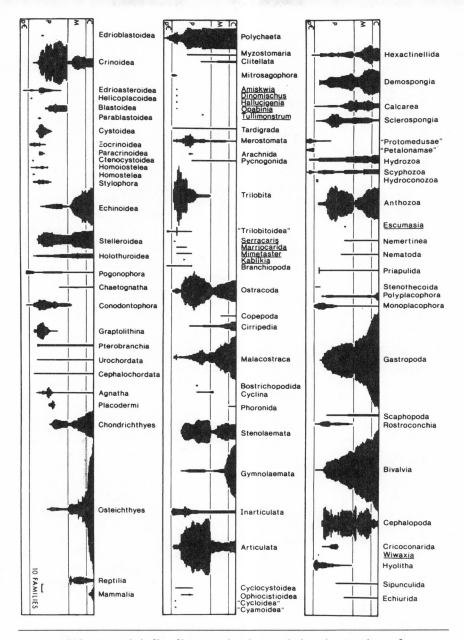

FIG. 5.2. Spindle diagram showing variation in number of marine families within metazoan biological classes during the last 650 million years of geologic time. Spindle width indicates number of families representing the class at any one time; scale is in lower left corner. In each strip, latest Precambrian (mid-Vendian) is at left (pЄ, at top of strip), followed to the right by successively younger Paleozoic (P), Mesozoic (M), and Cenozoic (C) eras. Width of these time intervals is proportional to duration. Mammalia through Agnatha (lower left) are vertebrate classes, and remaining units are those many-celled invertebrate animal groups for which there is a fossil record. From J. J. Sepkoski, Jr., "A Factor Analytic Description of the Phanerozoic Marine Fossil Record," *Paleobiology* 7(1981):36–53, by permission of the author and the Paleontological Society.

tarian approaches to interpreting earth history are said to be invalidated by evidence from the geologic record. Creationists say that "the present is *not* the key to the past," therefore they are "warranted in considering catastrophism as an alternative interpretation of the geologic strata. . . . There is no type of geologic feature which cannot be explained in terms of rapid formation."[11] "This does not mean that the basic laws of nature were changed but that the rates of geologic and other processes were profoundly disturbed and altered."[12]

2. "The creation model . . . predicts" that virtually all sedimentary rocks "must have been deposited in one single great epoch of deposition" during "a universal hydraulic cataclysm."[13] "Most geologic formations are of such character as to require natural forces operating far more intensely than is found in the modern world."[14] Terrigenous clastic sediments (such as sand and clay) were supposedly deposited rapidly and then lithified (transformed into stone) by extremely rapid precipitation of cements. According to the creationists, the formation of limestones, dolomites, and evaporites is best explained "in the context of a hydraulic cataclysm" in which rapid and "massive precipitation from solution in chemical-rich waters" occurred.[15] In addition, coal and oil were "almost certainly formed by some kind of catastrophic burial."[16]

3. Creationists claim that igneous rocks "were formed rapidly,"[17] and that the cooling and crystallization of igneous intrusions such as granitic batholiths is a rapid process. Since they believe the formation of igneous rocks is "rapid and of short duration," they say "the same must likewise be true of metallic flows [ore deposits] associated with them."[18]

4. "All the organisms of the fossil record were originally created contemporaneously by the Creator during the creation period. Thus they lived together in the same world."[19] "Rightly interpreted, the fossil record *does* show that all the forms of life in the fossils existed contemporaneously from the beginning."[20] "The very existence of fossils necessarily speaks of rapidity of formation! Fossils are not produced by slow uniformitarian rates of sediment deposition . . . it should be obvious that the actual formation of potential fossils . . . requires rapid and compact burial of the organisms concerned, and this requires catastrophism."[21] "The vast deposits of fossils testify unequivocally to rapid, catastrophic burial."[22] "Fossilization is unnatural, abnormal, catastrophic, quick, unique, exceptional, cataclysmic."[23] "The vast fossil record . . . is . . . a record . . . of the sudden destruction of life. It must have formed sometime after man first sinned, and therefore must have been formed cataclysmically, not gradually."[24]

5. The validity of biostratigraphy, the science that attempts to determine the relative ages of sedimentary strata based on fossil content, is rejected categorically by creationists. They claim that the arrangement of the geologic time scale, as constructed by biostratigraphers and other geologists, "is based on the assumption of evolution."[25] "Since the arrangement of fossils [in temporal succession] is arbitrary, i.e., based upon the *a priori* assumption of evolution, then the geologic column cannot be used to demonstrate evolution or vast age."[26] "Although the fossil record has been interpreted to teach evolution, the record itself has been based on the assumption of evolution. The message is a mere tautology. The fossils speak of evolution, because they have been made

to speak of evolution. . . . There is no truly objective time sequence to the fossil record, since the time connections are based on the evolutionary assumption."[27] Thus the creationists claim that the construction of the geologic column is "obviously a powerful system of circular reasoning."[28]

6. Creationists cite two general lines of evidence to prove that geologists have misinterpreted the geologic column. First, creationists argue that "out-of-order" rock sequences in which older rocks occur above younger rocks (as dated by geologists)—as in the Lewis Overthrust of Montana—indicate that the sequence of geologic ages has no objective basis. "It is quite possible for any vertical sequence of the 'ages' to exist in any given locality. Any age may be on the bottom, any on the top, and any combination in between."[29] Second, the supposed co-occurrence of fossils from different geologic ages in the same stratum—for example, human and dinosaur footprints in Texas and the presence of pine pollen in trilobite-bearing Cambrian shales in the Grand Canyon—is also taken as evidence that geologists' view of the geologic column is not objective.

7. Creationists admit, however, that a "general order from simple to complex" characterizes "the fossil record in the geologic column," but they suggest that the creation model explains this phenomenon "with more precision and detail" than other scientific models.[30] The general sequence of "distinctive fossil assemblages found in different rock systems" is "believed to be attributable to a combination of ecologic and hydraulic factors."[31] "The different geologic 'ages' are actually different ecologic zones in one antediluvian age."[32] Hydrodynamic factors in the cataclysmic flood milieu would supposedly sort the organisms "into assemblages of similar size and shape."[33] Morris has further explained:

> In any one locality, of course, there would be a definite tendency for similar kinds of animals to be buried at about the same level, and for different kinds to be buried in increasing size and complexity. Simple marine organisms would be buried first, then fishes, then amphibians, then reptiles, birds and mammals. This is in order of (1) increasing elevation of natural habitat; (2) increasing hydrodynamic resistance to gravitational settling in the sediment-bearing waters; and (3) increasing ability to flee from the encroaching floodwaters. This is exactly what is commonly found in sedimentary rocks.[34]

8. Creationists claim that the theoretical basis for radiometric age dating, as used by most modern geologists, is completely erroneous: "All the uranium-lead measurements, the potassium-argon measurements and all similar [radiometric] measurements which have shown" the earth to be billions of years old "have somehow been misinterpreted."[35] Furthermore, "it appears that geochronology is an example of a mixture of assumptions, guesses and imposed imagined universal principles."[36] "Most creationists . . . have viewed the evidence regarding the age of the earth as pointing to a very young age of from about 7,000 to 10,000 years. They maintain that not only do various physical indicators, not radiometric in nature, but also the radiometric indicators themselves, point strongly to a young earth."[37] Creationists claim that since creation

there has been a continuous exponential decay of the earth's magnetic field, which demonstrates conclusively that the earth cannot be more than about ten thousand years old. Although they believe the earth is very young, they think it was initially created with a "superficial appearance of age."[38] "It was necessary for the world to be created with an appearance of age. The 'apparent age' at the time at which the cosmic clock was set when it was first wound up, was evidently tremendously large, perhaps to emphasize the eternity and transcendence of the Creator."[39]

9. Two general sources of water are said to have contributed to the global hydraulic cataclysm: (1) a vast downpouring of water from the skies, and (2) a vast outpouring of water from the earth's crust. Not coincidentally, these are the same two sources identified in Genesis 7:11. Creationists claim that in antediluvian times "there was no great or deep ocean" and "no deserts or ice-caps."[40] Most creationists believe that the antediluvian world was shrouded in a "water vapor canopy," and that the accompanying "greenhouse effect" allowed the earth to have warm, equable climates from the equator to the poles. This "water canopy" supposedly precipitated in a great downpour during the global hydraulic cataclysm.[41] The creationists further propose that "great volcanic and tectonic movements" accompanied the global hydraulic cataclysm.[42] "Great volcanic lava flows, earth movements, violent windstorms, and other catastrophes, including the great Ice Age, were after-effects of the Flood."[43] "After the Flood the earth gradually adjusted itself to new land and sea balances. Geological processes gradually approached their present rates and natural phenomena in general finally became more or less stabilized with their present characteristics."[44]

Perhaps the most concise statement of the creationist viewpoint concerning the geologic processes that occurred during the global cataclysm, which is invoked to explain most of the geologic record, appears in the book *Scientific Creationism,* prepared by the Institute for Creation Research:

> [It began with] a great hydraulic cataclysm bursting upon [the world] with currents of waters pouring perpetually from the skies and erupting continuously from the earth's crust, all over the world, for weeks on end, until the entire globe was submerged, accompanied by outpourings of magma from the mantle, gigantic earth movements, landslides, tsunamis, and explosions . . . all land animals would perish. Many, but not all, marine animals would perish. . . . Soils would soon erode away and trees and plants be uprooted and carried down toward the sea in great mats of flooding streams. Eventually the hills and mountains themselves would disintegrate and flow downstream in great landslides and turbidity currents. . . . Vast seas of mud and rock would flow downriver, trapping many animals and rafting great masses of plants with them. On the ocean bottom, upwelling sediments and subterranean waters and magmas would entomb hordes of invertebrates. The waters would undergo rapid changes in heat and salinity, great slurries would form, and immense amounts of chemicals would be dissolved and dispersed throughout the seaways. Eventually, the land sediments and waters would commingle with those in the ocean. Finally the sediments would settle out as the waters slowed down,

dissolved chemicals would precipitate out at times and places where the salinity and temperature permitted, and great beds of sediment, soon to be cemented into rock, would be formed all over the world.[45]

The creationists have indeed postulated a relatively detailed account of earth history. In many respects, because it is so detailed and all encompassing, creationist geology is the most vulnerable to scientific scrutiny of their various positions. Almost all modern-day geologists find the creationist model untenable in virtually every respect. The creationists are correct in pointing out, however, that truth in science is not necessarily determined by majority rule.

SCIENCE VERSUS CREATIONISM

The creationist and scientific interpretations of geologic history are now summarized side by side, the scientific view appearing first since it represents the virtually unanimous consensus of geologists and other scientists working with the relevant data. The two competing models are then evaluated.

SCIENTIFIC

1. Earth history can be interpreted through study of the rocks of the earth's crust, assuming only the invariance of the basic physicochemical laws of nature.

2. Most geological processes act slowly by human standards, although sporadic violence as in storms or volcanic eruptions is normal.

3. The age of the earth is about five billion years, and the sedimentary succession, with its detailed history of physical and biological evolution, accumulated progressively over an interval of greater than three billion years.

4. The stratigraphic record preserves samples of life forms that existed individually for relatively brief periods through three billion years. It is characterized by originations and extinctions at all biological levels, from the species to the class or phylum. Almost all biological groups display cumulative change if traced for significant intervals of time. That is, they evolve.

CREATIONIST

1. *"Any real knowledge of origins or of earth history antecedent to human historical records can only be obtained through divine revelation."*[46]

2. *Earth history has been dominated by catastrophe. "There is no type of geologic feature which cannot be explained in terms of rapid formation."*[47]

3. *"The earth is really only several thousand years old, as the Bible teaches."*[48] *Virtually all sedimentary rocks must have been deposited in one single epoch of deposition during a universal hydraulic cataclysm.*

4. *"All the organisms of the fossil record were originally created contemporaneously by the Creator during the creation period. Thus they lived together in the same world."*[49] *The "general order from simple to complex" that characterizes "the fossil record in the geologic column"*[50] *is "attributable to a combination of ecologic and hydraulic factors."*[51]

EVALUATING THE MODELS

We now turn to the principles and data that are most important in choosing the creationist or scientific interpretations of earth history.

UNIFORMITARIANISM

Uniformitarianism, as understood today and used herein, is based on the assumption that the physicochemical laws of nature do not vary in space and time. This is the fundamental principle of geology and indeed of science as a whole. Because the validity of this principle is denied by creationists, we should briefly examine the history and present status of the doctrine.

Although uniformitarianism is implicit in the writings of fifteenth- and sixteenth-century naturalists such as Leonardo da Vinci,[52] the concept was first enunciated by James Hutton in 1785 and subsequently elucidated by Charles Lyell (1830–1833) in his *Principles of Geology*. Lyell postulated a dynamic earth, changing progressively in accordance with the invariance of the laws of nature. His arguments were immediately compelling, their effects far-reaching; within a short time, uniformitarianism was accepted generally and replaced the catastrophist model that had so strongly influenced earlier interpretations of earth history. The effect was to release geology from its function as "the efficient Auxiliary and Handmaiden of Religion"[53] and to permit it to assume its appropriate position in science.

From its inception, the doctrine of uniformitarianism has involved two concepts.[54] This duality has been the basis—because it has been unrecognized—for much of the confusion surrounding the doctrine. Lyell's first and basic assertion was that the laws of nature have remained invariant in space and time; this assumption is termed *methodologic uniformitarianism,* and because it is consistent with available scientific data, it provides the conceptual basis for much of science. This assumption enables geologists to interpret the physical and biological history of the earth by comparing the rocks and fossils of the geologic record with field and laboratory observation of the products of modern processes. Creationists deny the applicability of uniformitarian principles to interpretation of most of the geologic record, especially as it relates to events that supposedly occurred during the biblical flood or during the formation of the earth and its life forms.[55] Instead, they suggest that episodes of supernatural intervention (primarily during the creation period and the biblical flood) marked periods of "profound discontinuity in the normal processes of nature."[56]

In addition to postulating that natural forces operating over long periods of geologic time produced slow cumulative change, Lyell maintained that the rates of change have been more or less constant. This *substantive uniformitarianism* is a testable theory that was soon proved invalid, since the rates of geologic processes may change with time. For example, the rate of origination of the main invertebrate groups was unusually high in the early Paleozoic, whereas the rate of extinction was exceptionally high in the late Paleozoic (Figure 5.2). Similarly, the rates of erosion prior to the appearance of land vegetation must have been greater than they were subsequently. In both of these cases, rates of geological processes have changed without the necessity of change in the basic biological and physicochemical laws.

Interestingly, some creationists apparently accept methodologic uniformitarianism as a valid concept; they "do not question the assumption of uniformity of the basic laws of physics," at least subsequent to the "deluge," but "only question the assumption of uniformity of rates of geological and other proc-esses."[57] Because geologists also question substantive uniformitarian assump-tions, creationist criticism of the principle of uniformitarianism is, in part, a straw man.

What is more important, however, is that creationists have portrayed uniformitarianism in a grossly inaccurate manner. For example, Morris has writ-ten, "The principle of uniformity . . . explicitly denies any great geologic catastrophe as the Flood."[58] This is not so. There is no a priori assertion in-herent in methodologic uniformitarianism that prohibits geologic catastrophes from occurring; in fact, many catastrophic events have left their mark on the face of the earth — a number have occurred in our day (for example, the erup-tion of Mount St. Helens). Geologists do maintain, however, that catastrophic events can be understood in terms of the basic laws of nature and by the record those events have left in the earth's crust.

The creationist literature is also replete with claims that the existence of certain ancient sedimentary formations invalidates uniformitarianism because no identical counterparts are forming in the modern world. For example, citing the Shinarump Conglomerate of the Grand Canyon region, Morris argues that "nothing like this is being formed in the world today, as uniformitarianism should require."[59] Whitcomb and Morris also claim that "to be consistent with uniformitarianism the various types of sedimentary rocks must all be inter-preted in terms of so-called environments of deposition exactly equivalent to present-day situations."[60] These views of uniformitarianism are categorically wrong; geologists recognize many ancient sedimentary regimes not present in the modern world (for example, those that produced widespread black shale, banded iron formation, or thick widespread evaporites). These examples are evidence of a dynamic earth with a complex history, but they certainly do not demonstrate that methodologic uniformitarianism is false.

Even though some creationists ostensibly accept methodologic uniform-itarianism, their literature expresses an abhorrence for the use of uniformitarian principles in interpreting earth history. Uniformitarianism is viewed as a serious threat to biblical fundamentalism, which is the real basis of scientific crea-tionism. In 1951, E. J. Carnell wrote of the necessity to "rescue Christianity from the jaws of science once the principle of uniformity destroys God's right to perform miracles."[61] Other creationists have argued that "Christians who ac-cept by faith the historicity and inerrancy of the Book of Genesis . . . have no reason to fear that scientific discovery will contradict the Scriptures."[62] Henry Morris has even claimed that the principle of uniformitarianism fulfills the prophecy recorded in 2 Peter 3:3-4, where it is written that some men in the "last days" will ridicule Christian teachings by saying that "all things con-tinue as they were since the beginning of creation."[63] Morris has written, "The philosophical foundation of this denial [of Christian teaching] would be the principle of uniformity!"[64] However, uniformitarian approaches to interpreta-tion of earth history have shown clearly that all things have not continued "as they were."

The creationist rejection of uniformitarian principles in interpreting earth history undoubtedly stems from the firm belief that supernatural forces were responsible for creating most of the geologic record. Significantly, the creationists have failed to note the basic distinction between methodologic and substantive uniformitarianism and have therefore misrepresented the geologists' position on the subject.

If methodologic uniformitarianism is acceptable, the details of earth history can be interpreted through the formulation of an endless series of hypotheses that can be supported or rejected with factual data. If uniformitarianism is denied, then either we must abandon any attempt to decipher earth history or we must depend on divine revelation, which is clearly outside the realm of scientific inquiry.

SEDIMENTARY ROCK RECORD

Sedimentary deposits that were formed rapidly or catastrophically have characteristics that differentiate them from those deposited more gradually. Because sedimentary processes, whether gradual or catastrophic, obey the basic laws of physics and chemistry, ancient sedimentation processes can be understood by applying the principle of uniformitarianism to the rocks containing a record of the physical, chemical, and biological processes of sedimentation. In this way, interpretations of ancient depositional environments can be tested scientifically in the laboratory and the field.

The creationist claim of global catastrophism, however, is unsubstantiated for even a small portion of the geologic column, much less the entire sedimentary record. Even though little if any geologic evidence has been gathered by the creationists to support their position, they insist that all the

> implications of diluvial catastrophism are borne out by the actual character of the sedimentary rocks and their contained fossils. Details may be difficult to decipher at particular locations, but this basic framework is an adequate interpretative tool, and the difficulties of detail are far less serious than those entailed in the evolutionary and uniformitarian framework.[65]

Geologists, on the other hand, argue that the very details of the sedimentary record refute the creationist contention that the entire sedimentary rock record was deposited catastrophically, not because they have been blinded by an inflexible doctrine of substantive uniformitarianism, as some creationists suggest, but because the great bulk of sedimentary rocks preserves features inconsistent with catastrophic deposition. The number of technical deficiencies in the creationist interpretation of sedimentary rocks is enormous; here we focus on only a few of the main weaknesses.

Because the creationist model is so all encompassing, it is especially vulnerable to negative evidence. Creationists carefully choose only those examples from the sedimentary record that may contain some evidence of "catastrophic" deposition. Because their model explicitly states that the *entire* geologic column was deposited catastrophically, any example that is inconsistent with cataclysmic deposition disproves their entire thesis. In contrast, the uniformitarian approach is not disputed by the discovery of any evidence of "catastrophic" deposition within the geologic column.

"Polystrate trees" serve as an example of the kind of limited evidence that creationists use to support their position.[66] According to creationist reasoning, the roots and basal trunk of some fossil trees at a few localities are preserved in vertical (life) position and penetrate through several feet of layered strata; the sediment enclosing the trees was deposited rapidly, or else the trees would have rotted before they were buried completely; therefore, the entire geologic record was deposited cataclysmically. Although geologists recognize that rapid sedimentation occurs sporadically at diverse sites and that "polystrate trees" could be anticipated in some lake or floodplain environments, such an occurrence, they argue, does not confirm the creationist claim of universal catastrophe because of the general characteristics of sedimentary rocks.

Sedimentary rocks can be divided into several major categories. One of the largest comprises the clastic rocks, the components of which were derived from the weathering of preexisting rocks, usually on land (terrigenous). Discrete particles produced by weathering are transported by water, wind, or ice to an area of deposition. Compaction or cementation later transforms these unconsolidated sediments into rock. The commonest terrigenous clastic rocks include sandstone (consisting largely of quartz sand) and shale (mostly clay). Modern terrigenous clastic rocks are deposited in a wide range of environments, which are actively studied by sedimentologists. Sedimentation experiments—in which, for example, sedimentation rates are allowed to vary—provide additional evidence that helps to explain depositional systems. Geologists attempt to relate this information, as well as a host of other geologic data (for example, petrographic, paleontologic, stratigraphic), about the modern physical processes of sedimentation, to the evidence of sedimentation recorded in the rocks.

What have tens of thousands of geologists found? The geologic column contains numerous sedimentary structures and textures that are indistinguishable from those forming in modern environments. In many cases the ancient record clearly contains abundant evidence of ancient fluvial, or river, systems. River channel, floodplain, overbank, levee, swamp, and other fluvial deposits are common, and the geometry and stratigraphic relations of these deposits are indistinguishable from those observed in modern environments. Where fluvial systems merge into marine deposits, well-defined deltaic systems are recognized. The geometry, stratigraphy, and sedimentary structures of many of these ancient deltas are indistinguishable from those forming at the present time.

Why, then, should geologists eschew the most straightforward and simple explanation that remains consistent with all the data and consider the deposits that look just like modern fluvial and deltaic systems to be the result of some short-lived global hydraulic cataclysm? The question that creationists should be asked is, How were deposits that look exactly like modern fluvial and deltaic systems formed in the midst of a global hydraulic cataclysm?

Another serious problem with the creationist model is that it overlooks the evidence of glacial deposits. Although creationists acknowledge that an ice age occurred after the Flood, they have largely ignored or denied the evidence of earlier ice ages as indicated by the widespread occurrence of ancient glacial deposits throughout the geologic column. Because of this omission, the creationist model is able to predict that virtually the entire sedimentary record was deposited in a short-lived global hydraulic cataclysm and that the antediluvian world was characterized by warm climates from equator to poles. However, geologists have found a variety of unique sedimentary deposits, especially tillites and striated pavements, that provide evidence of ancient glacial environments. Such glacial deposits have been identified in Late Precambrian, Late Ordovician–Early Silurian, Carboniferous-Permian,[67] and Quaternary strata. The problems these deposits pose for the creationist model are obvious. How did glacial deposits form *during* the Flood, for example, a time when the entire earth was supposedly covered with water as a result of a cataclysmic event?

The creationist response is that these ancient glacial deposits were merely formed by poorly understood processes within the Flood milieu. Consider Whitcomb and Morris's explanation of the widespread Permo-Carboniferous tillites:

> In such a geologic cataclysm as the Bible describes the Deluge to be, it is easy to visualize the possibility of some great volcanic or turbidity current type of phenomenon centered over the southern hemisphere . . . without any glacial action necessary at all. . . . By far the most reasonable way of understanding such deposits as these would be in terms of catastrophic diluvial action, with currents flowing from different directions and containing different sediments.[68]

In fact, it is not "easy" to visualize such processes forming tillites, whereas it is quite natural to suggest that glacial processes were responsible for their formation. After all, these ancient deposits have the same characteristic features that are being formed by glaciers today.

Creationists also have difficulty interpreting carbonate rocks. This group forms another major category of sedimentary rocks, of which limestone ($CaCO_3$) and dolomite ($CaMg\,[CO_3]_2$) are the most common. Creationists assert that "nothing less than massive precipitation from solution in chemical-rich waters [during the hydraulic cataclysm] . . . seems adequate to account for "the formation of limestones.[69] Concerning the origin of dolomite, they assert that "it seems that only direct precipitation from magnesium-rich flood waters can explain them."[70]

Does any evidence bear out these unreserved claims? Creationists cite no evidence, but abundant evidence contradicting these claims can be found in the geologic literature. Most limestones are not formed by massive precipitation from solution; rather, most "carbonate formation is basically biochemical and organisms are all-important in creating and modifying all types of carbonate particles."[71] In addition, most dolomite in the geologic record is not a product of direct and rapid precipitation, but is a replacement product of limestone. Although inorganic precipitation is recognized as a significant means of carbonate deposition, the abundant skeletal carbonate grains in most limestones indicate that organisms are important in the formation of limestone. Perhaps the Cretaceous chalks of North America and Europe are among the most striking examples. These chalks are composed of trillions of microscopic calcareous fossils, primarily planktonic algae and foraminifers. Other limestones are formed by the in situ accumulation of skeletal grains (for example, shells) derived from successive generations of lime-secreting benthic organisms. Some limestones preserve fossil corals and shells of other organisms in life position, so that fossil coral reefs and other types of organic carbonate buildups are common features in the geologic record. The sedimentologic and paleontologic evidence preserved in both ancient and modern reefs indicates that they developed gradually by accumulation of successive generations of lime-secreting organisms. The mere presence of fossil reefs in the geologic record should be an embarrassment to the creationists who are proponents of cataclysmic deposition. However, they merely dismiss the inconsistencies revealed by such deposits. For example, Whitcomb and Morris have denied that fossil reefs pose any problems:

> Particularly during the Flood, the extensive reefs formed in the warm waters of the antediluvian seas would have been eroded and re-deposited, often giving the appearance now of an ancient reef of great extent. In any case, it is evident that it is possible to explain coral reef formation, whether ancient or modern, in terms of Biblical geochronology.[72]

D. B. D'Armond's explanation of the formation of the famous Silurian carbonate buildups or "reefs" of the midwestern United States is also interesting:

> Following the mid-Flood deposition attributed to tidal effects, a rapid emergence of continental land masses started to occur, triggering additional violent crustal movements which, in turn, caused large numbers of tsunamis to sweep over newly emergent shorelines. The newly formed Silurian deposits, being uplifted, became a shoreline area capable of receiving coral reef fragments torn loose and transported by tsunami-type waves. The source area for these reef materials could have been actual antediluvian reefs growing on Precambrian basement rocks in the general area of present-day Hudson Bay.[73]

Thus by dismissing evidence for in situ growth, creationists can simply postulate that the reefs had already formed during the antediluvial period and were transported piecemeal and then reconstituted many hundreds or thousands of miles distant during the hydraulic cataclysm.

The earth's crust contains a tremendous quantity of carbonate rock, probably around 10^{23} grams.[74] If, as creationists believe, virtually this entire mass was precipitated from solution during the diluvial cataclysm, the initial floodwater solution necessary to accomplish this task can be estimated by supposing that this mass of carbonate was dissolved in the total quantity of water available on the earth's surface (about 1.5×10^{23} grams). The result would be a solution of about 650 grams of calcite and dolomite in every liter of water. Calcite is more soluble than dolomite, but calcite solubilities even under the most ideal circumstances are not known to exceed several tenths of a gram per liter.[75] This means that creationists must devise a means to increase either carbonate solubilities or the total volume of water by several orders of magnitude. The creationist hypothesis also ignores the fact that these same floodwater solutions must have carried trillions of tons of other dissolved materials, including silica, iron sulfides and oxides, phosphate, and evaporite salts.

Evaporites, another group of sedimentary rocks, are composed primarily of highly soluble mineral salts. Common examples include halite (rock salt) and gypsum. Owing to the high solubility of these salts, concentrations considerably greater than those of normal seawater are necessary before precipitation can occur. For example, calcium sulfate concentrations greater than three times those of normal seawater are required for precipitation of gypsum, and a tenfold concentration is required to produce halite.[76] Magnesium and potash salts require concentrations twenty or more times that of seawater before precipitation begins. Not surprisingly, modern depositional environments for evaporites are in climates of high net evaporation. As water evaporates, the dissolved salts are concentrated progressively to the level at which precipitation occurs. Many ancient evaporite-bearing sedimentary sequences closely resemble those forming in modern evaporite environments, such as playa lakes and the sabkhas and adjacent areas of the Persian Gulf. Thick evaporites in the geologic record are characteristically restricted to enclosed structural basins, where, presumably, the net inflow of marine waters to the basin was less than the net evaporation.

The precipitation of evaporite minerals during a diluvial cataclysm should pose insurmountable problems for the creationist. If the floodwaters covered the earth, why are evaporites found only in certain specific areas? How did the floodwaters achieve local concentrations of dissolved salts high enough for precipitation to begin? What was the source of the dissolved salts? In response to these questions, some creationists have suggested that brines of different composition were mixed in the cataclysmic milieu and that the result was massive precipitation.[77] Whitcomb and Morris have suggested some additional possibilities:

> there is always the possibility that the evaporite bed was formed by transportation from some previous location, where it may have existed since the Creation. And there is also the possibility that it may have been formed by intense application of heat for evaporation of large quantities of water in a short time rather than ordinary solar heat acting over a long time.[78]

Some creationists have further asserted that "evaporite beds actually constitute a serious problem to the uniformitarian model. There is no present-day process at all capable of producing such formations. Evaporites clearly favor the cataclysmic model."[79] Such unabashed claims are presented without any observational basis.

Creationists also believe that the great accumulations of organic carbon preserved in the earth's crust, primarily coal and oil, were formed cataclysmically during the Flood. They have suggested that the coal beds of the world represent the preserved remains of huge masses of vegetation that rafted out on the surging floodwaters and were then deposited with other flood-derived sediments. They maintain "that the flood model of coal vegetation accumulation is much more realistic" than the conventional geologic interpretations.[80] The creationists have also stated that "oil is not being formed today, nor is it found even in Pleistocene deposits."[81] Petroleum geologists disagree with both parts of this statement. Although physical evidence invalidates the creationists' claims, a simple statistic can serve for rebuttal. Bolin has estimated that the amount of "carbon locked in coal and oil exceeds by a factor of about 50 times the amount of carbon in all living organisms."[82] Even under the improbable assumption that all organic carbon in the antediluvian world was preserved as coal or oil without being oxidized, the creationists must propose a preflood world with fifty times the biomass of the modern world. Both their proposal and their assumption are beyond credulity.

The contemporary flood environments envisaged by the creationists are just as difficult to accept scientifically. We are asked to believe that during a single global flood some parts of the earth were subjected to conditions that produced sediments that are identical with glacial tills, while in other areas there developed vast fluvial-deltaic systems, massive limestones, voluminous salt deposits, transported and reconstituted coral reef masses, and diverse fine sedimentary rocks that preserve the minute structural detail of their contained fossils. Furthermore, each of these rock types is purported to have been deposited successively in one locality within the few days' duration of the biblical flood. It would seem that the creationists are merely attempting to construct a story that remains consistent with their religious dogma.

BIOSTRATIGRAPHY AND THE GEOLOGIC TIME SCALE

Creationist literature rejects the geologic time scale, the yardstick against which earth history is measured, because it is "based on the assumption of evolution."[83] To evaluate this assertion we need to examine the dates when the major intervals of the geologic time scale were first proposed by scientists (see Table 5.1).[84]

With the exception of the Ordovician (established later to resolve a dispute over the boundary between Cambrian and Silurian), all intervals were proposed prior to publication of Darwin's *Origin of Species* in 1859. The Tertiary preceded the *Origin* by a full century. Components of the time scale were established on the basis of objective data, primarily superposition (observed stratigraphic succession) and fosssil content. The "wonderful order and regularity with which nature has disposed of these singular productions (fossils) and assigned to each its class and peculiar stratum" were recognized as early as 1796

TABLE 5.1. DATES WHEN GEOLOGIC TIME SCALES WERE FIRST PROPOSED

Period	Date	Era	Date
Quaternary	1829	Cenozoic	1841
Tertiary	1759		
Cretaceous	1822	Mesozoic	1841
Jurassic	1795		
Triassic	1834		
Permian	1841	Paleozoic	1838
Carboniferous	1822		
Devonian	1837		
Silurian	1835		
Ordovician	1879		
Cambrian	1835		

by William Smith, the father of stratigraphy. At about the same time, the noted French biologists Georges Cuvier and Alexandre Brongniart arrived independently at comparable conclusions. Cuvier noted that "these fossils are generally the same in corresponding beds, and present tolerably marked differences of species from one group of beds to another. It is a method of recognition which up to the present time has never deceived us."[85]

Both the stratigraphic column and our general understanding of fossil successions were thus developed empirically and without any reference to evolutionary assumption. In fact, Cuvier was a leading catastrophist who emphatically denied that evolution had occurred. The sequence of geologic periods was reasonably well established in 1841. It was not until several decades later that scientists rather suddenly began to recognize the evolutionary thread connecting successive assemblages of fossils.

THE FOSSIL RECORD

The creationist literature contains many false assumptions about fossils, the record of former life preserved in the earth's sedimentary rocks. In this section we address some contentions that bear on the creation-evolution debate.

Fossilization. Consider first the creationist misconception of the process of fossilization: "We have seen that the preservation of organic materials as fossils, *by whatever means,* requires some sort of catastrophic condition, some kind of quick burial by engulfing sediments, usually followed by some abnormal chemical means of rapid solidification. There is little wonder, then, that it is so difficult to find any remains of the modern era which could be said to be in the process of 'becoming' fossils."[86] On the contrary, remains of organisms are commonly incorporated as particles in the course of present-day accumulation of sediment layers. Limestones, in particular, may be composed largely of the remains of animals and plants. For example, marine green algae are the main source of the beds of unconsolidated lime mud that have formed in Florida Bay during the last eleven thousand years, which are known as the Holocene interval.[87] Sedimentary structures and radiometric dates show that accumulation has been slow, and abundant fossil snails, clams, and other organisms throughout the sequence represent elements of the fauna that lived at the time individual mud layers formed.[88] The fossils preserved within these layers are mainly the same species that inhabit modern environments in southern Florida. Most of the Holocene mud succession of Florida Bay is soft enough that it can still be penetrated fully with aluminum irrigation pipe, although the process of lithification is progressing at varying rates through compaction, desiccation, and cementation. Close analogs for these fossiliferous modern carbonates abound in the ancient geologic record in many areas of the world.

Origination and extinction. A second creationist misconception is that "all of the forms of life in the fossils existed contemporaneously from the beginning."[89] The sequence of fossils preserved in the sedimentary record, when examined at any biologic (taxonomic) level, clearly refutes this creationist claim. As Cuvier recognized over 150 years ago, the fossil record documents a series of changes in the history of life on earth, including extinctions and the appearance of new biologic groups. Scientists linked distinctive forms of life with distinctive periods of earth history long before the general acceptance of evolution.

Patterns of origination and extinction of biologic groups are best illustrated at the family level (see Figure 5.2). Originations and the resultant increases in diversity characterize the early Paleozoic (Cambrian through Middle Ordovician) and Early Mesozoic (Triassic and Jurassic), whereas extinctions predominated in the Late Devonian, Late Permian, Late Triassic, and Late Cretaceous. These patterns produce distinctive associations of organisms (Figure 5.3) that were recognized and named 150 years ago.[90] Paleozoic (Cam-

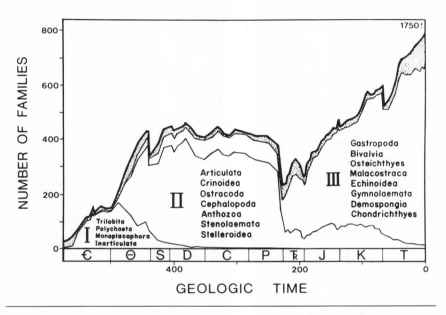

FIG. 5.3. Distribution of metazoan marine families during Phanerozoic time, the last 600 million years (data from Figure 5.2). Geologic time is represented by symbols for the Paleozoic (Є [Cambrian], Θ [Ordovician], S [Silurian], D [Devonian], C [Carboniferous], P [Permian]); Mesozoic (Tr [Triassic], J [Jurassic], K [Cretaceous]); and Cenozoic (T [Tertiary]). Width of these time intervals is proportional to duration. Dark uppermost curve is total families. Roman numerals correspond to adjacent biologic groupings, with family totals delimited by fine lines. Stippled field immediately below total diversity curve represents residual diversity not accommodated by these three biologic groupings. Number 1750 in the upper right corner is number of metazoan families described from modern oceans. Figure displays distinctive associations of fossils throughout the Phanerozoic. From Sepkoski, "A Factor Analytic Description of the Phanerozoic Marine Fossil Record," by permission of the author and the Paleontological Society.

brian through Permian) refers to ancient life, Mesozoic (Triassic through Cretaceous) to intermediate life, and Cenozoic (Tertiary and Quaternary) to recent life. However, new discoveries still refine the details of our knowledge; for example, the range of the earliest vertebrates has been extended recently from the Middle Ordovician to the Late Cambrian.[91] Nevertheless, the general patterns of origination and extinction are well established, and there is now little prospect of any great change in the overall picture.

Sequences of fossils and the Deluge. As noted earlier, creationists acknowledge that a general order exists in the stratigraphic succession of fossils, but they maintain that "all the organisms of the fossil record were originally created contemporaneously by the Creator during the creation period. They thus lived together in the same world."[92] Creationists also believe that "the fossil-bearing strata were apparently laid down in large measure during the Flood, with apparent sequences attributed not to evolution but rather to hydrodynamic selectivity, ecologic habitats, and differential mobility and strength of the various creatures."[93] Marine creatures were reputedly buried first, under highly selective hydrodynamic forces. They were followed by land plants and animals, those with the highest "mobility and strength" surviving longest before eventual burial in the youngest strata.

The fossil record does not support this model. If creationist contentions were valid, the oldest sediments would be expected to contain the greatest concentrations of sessile marine organisms (that is, ones that are permanently attached and therefore not free to move about). In fact, only single-celled algae and bacteria are present in the sedimentary record of almost three billion years that precedes the late Precambrian Vendian System.[94] In contrast, the Vendian contains nonskeletal metazoan animals, and the succeeding half-billion-year Phanerozoic is the one part of the record with a great diversity and abundance of heavy-shelled marine organisms that on hydrodynamic grounds could be expected to have settled first in the oldest of the "Deluge" sediments. Again, land animals and plants do not simply succeed assemblages of marine organisms, but are repeatedly and intricately interbedded with them. We might well ask whether the impressively huge carnivorous dinosaurs and other reptiles of the Mesozoic were weaker and less agile than the sheep and other grazing mammals that succeeded them in the Cenozoic. Were the Mesozoic fish somehow less capable of avoiding burial in the hydraulic cataclysm than the Cenozoic corals and snails that are found above them in stratigraphic succession? We must conclude that the similarity between the known distribution of fossils and the prediction of the creationist model is insufficient to provide a basis for serious comparison.

Evolutionary patterns. The fact of evolution—which refers to change with cumulative modification—has not been subject to informed challenge for about a century, and at least the general evolutionary paths for most major animals and plant groups have been known since the end of the nineteenth century. An example of the evidence for evolutionary change is presented in Figure 5.4.

Creationism does permit "diversification," but only within individual "kinds" ("Man Kind, Horse Kind, Dog Kind").[95] It is unclear how evolutionary successions such as those documented for the two ammonoid biological families in Figure 5.4 relate to the concept of kind. Nor have the creationists explained how such patterns of change could occur repeatedly in stratigraphic sequences that were supposedly deposited during a single global cataclysm.

Although the fact of evolution above the population level is not questioned in scientific circles, the patterns of change from one population to the next are the subject of detailed investigation and lively debate. The debate centers on the relative importance of two evolutionary models, gradualism and punctuated equilibria.[96] Evolutionary gradualism proposes that genetic change from one population to the next is slow and is expressed as progressive morphologic drift (in one or more characters) within a temporal succession of populations (that is, those occurring successively in a single profile); discontinuities in morphologic change are attributed to local gaps in the geologic record. Advocates of the punctuated equilibrium model claim that successive populations generally show little directional change, and that the fossil record is characterized by stability (*stasis*), punctuated by frequent abrupt changes in morphology (*punctuation*). They believe that the observed morphologic discontinuities have arisen owing to rapid evolution in small isolated populations (*peripheral isolates*) followed by migration to outcompete ancestral forms.

An example that combines gradual directional change in some morphologic features and relative stasis in others (Figure 5.5) has been provided by B. A. Malmgren and J. P. Kennett,[97] who have reported a Late Cenozoic evolutionary lineage comprising five species of the foraminiferal genus *Globorotalia*. The succession was studied in a single 208-meter limestone core from the submarine *Challenger Plateau* west of New Zealand. It represents an objective stratigraphic succession through 8.3 million years, from the Miocene to the present. Between forty and sixty specimens were picked from each of seventy-two levels within the core. The study involved almost 30,000 individual measurements, which were taken without reference to sample numbers or depth so as to assure unbiased treatment. Trends in six of the variables show coherent patterns, three of which are illustrated in Figure 5.5. Mean number of chambers and mean roundness of the periphery display gradualism. The percentage of individuals possessing a keel decreased abruptly during the Early Pliocene dividing the periods of stability.

Reworked fossils. Observation of current geologic processes reveals that, under exceptional circumstances, fossils may be released by erosion from ancient sediments, transported, and eventually redeposited to form a reworked or displaced suite. According to uniformitarian doctrine, this reworking can be expected in ancient sedimentary sequences, and indeed it does occur.[98] However, almost all fossils are altered drastically or destroyed completely in the weathering and transportation processes, so that displacement of most reworked fossils is readily recognizable.[99] In a few cases, some particularly resistant structures, especially the microscopic reproductive bodies of plants (for example, pollen), may show little physical evidence of reworking. Intimate knowledge of the nor-

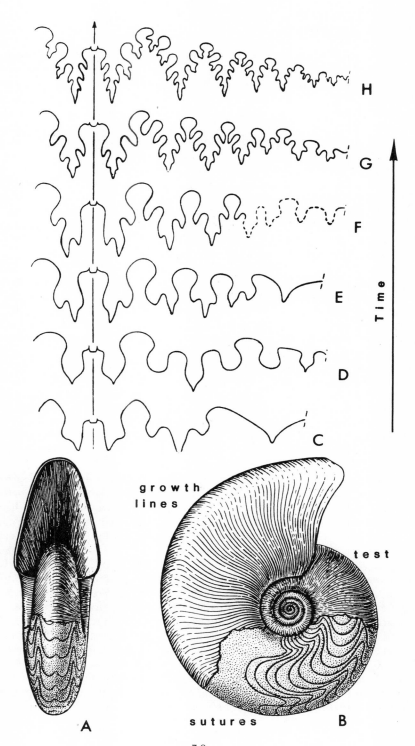

growth
lines

test

sutures

Time

H

G

F

E

D

C

A

B

◄ FIG. 5.4. Ammonoid cephalopods, extinct relatives of the living chambered *Nautilus,* provide an example of rapid evolution. *A, B,* mineralized conch, about natural size, with part of the outer shell removed to show sinuous margins of the transverse partitions or sutures. Marginal crenulation of the transverse partitions produced flexures in the sutural paths; progressive sutural change with geological time provides clear documentation of ammonoid evolution. *C–H,* mature sutures represent evolution of two ancestral-descendant ammonoid families through fifty million years, all enlarged. *C–E,* stratigraphic succession of the family Shumarditidae from the Upper Carboniferous of the Ural Mountains. *F–H,* stratigraphic succession of the family Perrinitidae from the succeeding Lower Permian of the southwestern United States. *A, B* modified from A. K. Miller and W. M. Furnish, "Paleozoic Ammonoidea," in *Treatise on Invertebrate Paleontology,* Part L, *Mollusca* 4, R. C. Moore, ed. (Boulder, Colo.; Lawrence, Kans.: Geological Society of America and University of Kansas Press, 1957), p. L13; *C–H* from V. E. Ruzhentsev, ed., *Fundamentals of Paleontology, Mollusca–Cephalopoda I.* (Moscow: Isdatel Akad. Nauk SSSR), p. 389. (In Russian.)

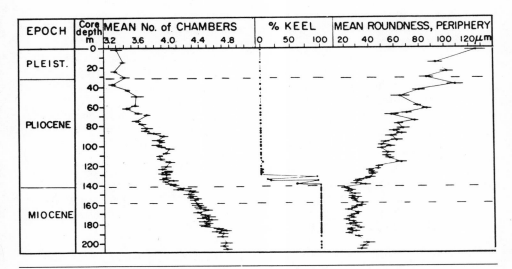

FIG. 5.5. Evolutionary patterns displayed by five species of the foraminiferal genus *Globorotalia.* Data derived from a single deep-sea core, 208 meters long, representing more or less continuous sedimentation for 8.3 million years, from the Miocene to the Recent. Horizontal lines are 95 percent confidence intervals. Data sets selected from B. A. Malmgren and J. P. Kennett, "Phyletic Gradualism in a Late Cenozoic Foraminiferal Lineage," *Paleobiology* 7(1981):230–40. Original study involved almost 30,000 individual morphologic measurements. Innumerable such sequences are available, but the immense effort involved in full treatment has discouraged adequate documentation in all but a few instances. Published by permission of the authors and the Paleontological Society.

mal biostratigraphic successions must then be employed to interpret the anomalous occurrences.[100] When such anomalies are encountered, they frequently lead to the recognition of subtle changes (for example, slight abrasion or color change) that confirm displacement. Such recognition may in fact add detail to reconstruction of geological history, as it documents an unconformity and period of erosion.

Displacement of fossils is thus both rare and generally readily recognizable. Contrary to creationist claims, displacement is not a major problem in establishing the evolutionary succession of life.[101]

Precambrian fossils. Although our knowledge of Precambrian fossils has increased greatly over the past two decades, it is still poor compared with that for Cambrian and younger intervals. Part of the reason for this lack of information is that these oldest of organisms were microscopic in size. They also had simple structures, and none is known to have developed a mineralized protective or supportive skeleton. Evidence of life in the form of bacteria and blue-green algae has been detected in the early Archaean samples that date back 3.5 billion years, but there is still some reservation about the biogenicity of these forms.[102] Microscopic plants were undoubtedly present in the Proterozoic, and eukaryotic (nucleated) cells appeared 1 billion years ago.[103] It now seems probable that changes in diversity and composition of these planktonic floras will afford an independent evolutionary chronology for the Proterozoic.[104] Megascopic animals first appeared in the late Proterozoic Vendian (Ediacaran) period beginning about 650 million years ago, and groups of shelly fossils appeared successively in the Cambrian about 560 million years ago.[105] Diversification of organisms from the Vendian through the Middle Ordovician was rapid, presumably because of the crossing of one or more biologic or ecologic thresholds as well as expansion into underutilized environments. This diversification cannot be represented as an abrupt event because it took place over a period of 200 million years.

Creationist literature seldom refers to Precambrian fossils, presumably because many of the data have been reported only recently. Apparently unaware of the late Proterozoic faunas described by M. F. Glaessner, M. Wade, and others, D. T. Gish, for example, stated that "not a single, indisputable, multicellular fossil has ever been found in Precambrian rocks!"[106]

Human footprints. Creationist literature abounds with references to the purported frequent occurrence of human fossils in ancient sedimentary rocks.[107] The examples date back to the Cambrian, long before the documented appearance of our earliest mammalian ancestor. If there were any substance to these reports, they would refute evolutionary chronology and cast serious doubts on the entire stratigraphic methodology. Significantly, all the so-called human fossils are said to be footprints, features that readily invite imaginative interpretation. No single example of a pre-Cenozoic human bone or tooth has been validated. As for the persistent claim that giant human footprints are directly associated with dinosaur tracks in the Cretaceous sediments of the Paluxy River, Texas, a refreshingly cautious creationist has expressed skepticism about such a possibility.[108] B. Neufeld reported that most Paluxy field occurrences reputed to be human tracks strain the imagination when interpreted as tracks of any origin. Sections of several laboratory specimens with some resemblance to human footprints revealed that the fine sedimentary layers are truncated at the margins of the prints rather than being depressed beneath them. The human "tracks" had been artificially carved. The author quoted Paluxy residents as reporting that carving footprints for sale to collectors had been a cottage industry during the depression. Although John D. Morris, son of creationist Henry Morris and holder of a Ph.D. in engineering from a reputable university, had concluded that "man and dinosaur walked at the same time and in the same place,"[109] Neufeld stated that the Paluxy River area "does not provide good evidence for the past existence of giant man. Nor does it provide evidence for the co-existence of such man (or other large mammals) and the giant dinosaurs."[110]

We can conclude that not one of the reputed pre-Cenozoic occurrences of human fossils has been validated. The oldest undoubted fossil man (genus *Homo*) is from strata in East Africa that date back only 2.5 million years, which is long after the extinction of the dinosaurs.[111]

AGE OF THE EARTH

Most dating of sedimentary rock sequences in geology is based on the science of biostratigraphy. The dates determined by biostratigraphers are, however, *relative* age dates, or dates that establish the relative age relationships of a stratigraphic unit with respect to other units. Relative age dating techniques produce evidence of a rock unit's precise position in sequence within the geologic time scale but provide no direct evidence of the *absolute* age of the rock unit. Absolute ages are now determined largely by radiometric dating techniques, whereby ages in years before the present can be assigned, within statistical limits, to the time of formation of a particular rock unit. Since specific radioactive isotopes are known to decay into daughter elements at a particular rate, as described by a physical decay constant, absolute ages are determined by measuring the ratio of undecayed radioactive isotopes to their daughter elements in a rock sample. Half the original complement of radioactive nuclei will have decayed in a particular period of time known as the half-life. Dates in excess of 3.5 billion years have been determined radiometrically for earth rocks, and even older dates have been determined for moon rocks and meteorites.

Since creationists believe that the earth and universe are no older than six thousand to ten thousand years, it is not surprising that they reject the validity of all known radiometric dates as well as other evidence that confirms that the earth and universe are billions of years old. Creationist criticism of radiometric dating techniques is based primarily on the contention that, because initial concentrations of parent-daughter elements are unknown and later introduction of daughter element contamination is possible, radiometric dates are wildly erroneous.[112] Geologists are aware of these problems and allow for them in their dating procedures. Since several different radioactive decay series are commonly used in analyzing a rock sample, for example, radioactive dating actually provides several independent means of determining the age of the sample. When all radiometric dates from a sample are concordant, scientists consider the age to have a high degree of reliability. Creationists are not persuaded by concordant dates but instead maintain that geologists have erred in determining the age of the earth by four or five orders of magnitude. They have also questioned the validity of radiometric dating by suggesting that nuclear decay "constants" are not invariant but vary with time in response to external or intrinsic parameters.[113] S. G. Brush has examined the creationists' criticisms of radiometric dating and has concluded that "they are based on ignorance or misunderstanding."[114] The reader should consult Brush's report for a thorough treatment of this highly technical subject.

Another method of radiometric dating that does not depend on direct measurements of parent and daughter elements, namely fission-track dating, also verifies the great antiquity of the earth. Spontaneous fission of uranium atoms in a rock sample produces microscopic tracks, the density of which is related to the age of the sample and its original uranium content. By comparing the density of the fission track before and after it is irradiated with a known dose of neutrons, and with uranium decay constants known, the absolute age of the rock can be determined.[115]

Creationists claim "that there is no sound physical evidence that the earth is very old" and "radioactive decay processes" are "equally, if not more, consistent with a very short time span."[116] But what physical evidence do the creationists present to verify their contention of a young earth? They have suggested that atmospheric helium content, the thickness of meteoric dust, and the rate of influx of dissolved solids into the ocean are all evidence of a young earth.[117] These suggestions erroneously assume that helium has not escaped from the atmosphere into space, that meteoric dust has not been reworked into the great thickness of sediments in the earth's crust, and that material has not precipitated from the oceans but has remained in solution.

The supposed exponential decay of the earth's magnetic field proposed by T. G. Barnes is also frequently cited by creationists as proof that the earth is only around ten thousand years old.[118] However, this second "proof" is based on dubious assumptions. First, Barnes assumed that the earth has no internal mechanism for maintaining a magnetic field and that the magnetic field decays exponentially as its energy dissipates. Brush stated that "this assumption is flatly contradicted by almost all current research on geomagnetism."[119] Barnes determined the parameters of his exponential decay curve only on the basis of measurements of the earth's magnetic dipole between 1835 and 1965. From this limited data base he extrapolated the curve backward until an unbelievably large value for the earth's magnetic field was reached in 20,000 B.C. He concluded that the earth must, therefore, be younger than that date. Barnes's magnetic extrapolations are invalidated by a considerable body of data accumulated by geophysicists. Remanent magnetism preserved in the rocks of the earth's crust provides testimony of the history of the earth's magnetic field and shows that the field is not decaying exponentially: at times in the geologic past it was stronger than at present and at other times weaker. In addition, the polarity of the earth's magnetic field has reversed, back and forth, many dozens of times during the Phanerozoic.[120] These observations refute the exponential decay model of Barnes. Completely ignoring the large body of paleomagnetic data that contradicts his basic assumption, Barnes has simply extrapolated a very restricted data base all the way back to the origin of the earth.

Creationist and scientific interpretations of the history of the earth and its life forms differ fundamentally in every respect — in methodology, basic assumptions, and conclusions. Creationism employs a single model, derived primarily through literal interpretation of the Bible. Divine revelation provides the preeminent source of information, so that field and laboratory generation of new data is largely unnecessary. The geologic sciences employ the principle of multiple working hypotheses. Several alternative models are considered, and field and laboratory data are sought to support or refute the competing interpretations.

Creationism assumes that the Bible reveals the factual story of creation. Science assumes that the basic physicochemical laws of nature are invariant, so that geological processes occurring at the present time are a key to past earth history.

Creationists conclude that the earth is several thousand years old. The sedimentary record is the product of the Noachian Flood, and the fossil record results from progressive inundation and hydraulic sorting of animal and plant communities that were all contemporaries. Science claims that the age of the earth is approximately five billion years, that the physicochemical laws that control the processes acting at the present can account for the observed geological record, and that fossils illustrate the evolution of life over three billion years.

Both the creationist and scientific interpretations of earth history are based on assumptions. Our direct familiarity with rocks of the earth's crust and our studies of relevant scientific literature from many disciplines lead us to conclude that scientific interpretations are plausible and consistent with factual geologic data. Creationist interpretations are neither.

Geologists have no vested interest in forcing their data to fit any preconceived model. They are a restlessly inquisitive group and strive repeatedly to develop alternative models that better fit the data. In insisting on their literal interpretation of the Bible, creationists leave themselves with no alternative to the single creation model. Geologic data give them no support. Rather, the geologic data directly or indirectly refute virtually all "scientific" creationist assertions.

For Further Reading

BLATT, H.; MIDDLETON, G.; AND MURRAY, R. *Origin of Sedimentary Rocks,* 2nd ed. Englewood Cliffs, N.J.: Prentice-Hall, 1980.
  Summary of all major sedimentary rocks, their composition, and mechanisms of deposition.
COLBERT, E. H. *Evolution of the Vertebrates: A History of the Backboned Animals through Time,* 3rd ed. New York: John Wiley & Sons, 1980.
  A comprehensive account of vertebrate evolution.
DOTT, R. H., JR., AND BATTEN, R. L. *Evolution of the Earth,* 3rd ed. New York: McGraw-Hill, 1981.
  A basic textbook on the fundamentals of historical geology.

LAPORTE, L. F., ed. *Evolution and the Fossil Record.* San Francisco: W. H. Freeman, 1978.

A compilation of readings from *Scientific American* that provides an overview of the evolution and history of life as recorded by the sequence of fossils preserved in the earth's crust.

MCPHEE, J. *Basin and Range.* New York: Farrar, Straus & Giroux, 1981.

A highly readable account of the wanderings and ideas of several modern geologists.

RAUP, D. M., AND STANLEY, S. M. *Principles of Paleontology,* 2nd ed. San Francisco: W. H. Freeman, 1978.

An introduction to the nature of fossils and the uses of paleontological data.

RHODES, F. H. T., AND STONE, R. O., eds. *Language of the Earth.* New York: Pergamon Press, 1981.

A readable compilation that illustrates the scope and range of the geological sciences and conveys their flavor and style, rather than cataloging their content; displays all our knowledge as provisional rather than infallible, as refinable rather than complete and finished.

SKINNER, B. J., ed. *Paleontology and Paleoenvironments.* Los Altos, California: William Kaufmann, 1980.

A compilation of readings on invertebrate and vertebrate paleontology, from *Scientific American;* includes papers on evolutionary changes deduced from paleontological evidence.

[6]

The Origin of Life

JOHN H. WILSON

FROM the beginning of recorded history, mankind has wondered about the origins of the earth and heavens and, in particular, about the origin of living things. Early explanations, which vary widely from culture to culture, are stories we often call myths because they lack an observational basis and seem to us to be made up. In the last few centuries a new kind of explanation, a scientific explanation, has developed. Science may be defined as the observation, experimental investigation, and theoretical explanation of the phenomena of nature. All scientific explanations are only tentative answers to questions about nature. There are no absolute explanations in science, only working theories that must remain open to the demands of new observations and new theoretical insights. Thus, Ptolemy's earth-centered universe gave way successively to Copernicus's sun-centered universe and to the modern view of a sun-centered solar system on the outer edges of one smallish galaxy among billions in the universe. The continual revision of theories, in small or large degree, to reflect more accurately the observations of nature is the hallmark of science.

The scientific exploration of nature has revealed a few basic natural laws that govern the behavior of matter and energy everywhere on earth, throughout the solar system, and as far into the cosmos as we can see. That is not to say that we understand everything about the universe but rather that everything we understand occurs in accordance with these natural laws. At present, these laws are not seriously disputed by anyone and thus form a common ground for all scientific discourse. However, these laws have been derived from observations that span only a few hundred years, an instant in the time scale of the universe. In any search for origins, a critical question is, "How far into the past do these laws apply?"

The author wishes to acknowledge significant contributions by John Rodgers, Tom Gudewicz, and Lynda Thomas. Invaluable discussions with these individuals have helped to shape the viewpoints presented in this chapter.

The answer to this question provides the basis for categorizing various explanations of origins as natural, unnatural, or supernatural. Natural explanations posit that the laws of nature, as derived from our observations, have governed the behavior of matter and energy from the beginning of the universe to the present. Unnatural explanations posit that at some time in the past the laws of nature were different, perhaps radically different, from present laws and that the observable features of the universe can be understood only through these differences. Supernatural explanations posit that at some time in the past a sentient being, a creator, took an active part in shaping the universe. The terms natural, unnatural, and supernatural are used in this sense throughout this chapter.

Natural and supernatural explanations dominate popular accounts of origins. Evolution, which is the strongest natural explanation, holds that the gross features of the universe — including galaxies, solar systems, and planets; the transition from nonliving matter to living organisms; and the diversity of life forms, including human beings — all arose as a consequence of the innate proclivities of matter and energy, as expressed by the laws of nature. Creationism accounts for these same features of the universe by a number of supernatural acts whereby the universe was created in a fully functional form not very different from what we observe today. Although public attention has focused on these two alternatives, it is important to remember that they are not the only ones.

For creationists, then, life arose full-blown from the hands of the creator, and further mechanistic inquiry is beyond the realm of current science. By contrast, evolutionists demand from their theory a plausible mechanistic explanation for the transformation of nonliving matter into living organisms. This chapter presents their viewpoint in some detail and notes prominent creationist objections where appropriate. The evolutionary account presented here is but one version of several that are being actively discussed and experimentally tested. Undoubtedly, the evolutionists' explanation of life's origin will be revised — perhaps substantially — as knowledge accumulates. As the reader will come to appreciate, the evolutionists' account, although entirely plausible and consistent with natural laws, nonetheless contains several gaps in understanding that will require further experimental and theoretical insight if we are to explain them in detail.

HISTORICAL PERSPECTIVE

The scientific investigation into the origin of life is essentially a modern approach. Only a few hundred years ago most people, including scientists, were agreed that life arose from nonliving matter: mice were produced by rubbish, maggots by decaying meat, and microorganisms by broth. Ultimately, this concept of spontaneous generation, as it was commonly known, generated a scientific controversy that lasted more than two hundred years. Experimentalists chipped away at the idea until the 1860s, when Louis Pasteur erased common belief in spontaneous generation through a series of careful and convincing experiments. Even before the death knell of spontaneous generation, the new theories of Darwin and Wallace were attracting scientific attention to the possibility that life originated through an evolutionary process. However, with little biochemical information and no mechanistic theories, the scientific investigation into the origin of life lay dormant for more than fifty years. In the 1920s and 1930s the question was raised once again by the biochemists A. I. Oparin in Russia and J. B. S. Haldane in Britain. Both men singled out the probable absence of free oxygen in the earth's early atmosphere as a crucial factor that would favor a prebiotic (before life) chemical evolution. The insightful mechanistic speculations of these men form the foundation of much of modern scientific thought into the origin of life.

Scientists have only a broad general idea of the prebiotic conditions on the primitive earth. Although the earth is thought to be about 4.6 billion years old, there is no geologic evidence available for nearly the entire first billion years of its existence. Cellular life must have arisen early because microfossils of primitive, single-celled organisms are present in rocks that are 3–3.5 billion years old. These early life forms probably lived off organic molecules that had been synthesized by certain chemical processes, which are discussed later. Two to three billion years ago, according to the fossil record, organisms arose that could harvest energy directly from the sun. Oxygen, a waste product of these photosynthetic organisms, eventually poisoned the atmosphere for chemical synthesis. At the same time, it opened the way for efficient energy production through the controlled metabolic burning of organic fuels. The development of respiration in primitive organisms and the resulting glut of energy may have permitted the evolution of multicellular organisms, a change marked in the geologic record by an explosion of fossil forms about 600 million years ago. This brief history of the early evolution of life focuses attention on the very early stages in the earth's development as the time when life originated.

One underlying assumption of scientific inquiry is that life did originate on the earth. The possibility that life arose elsewhere in the universe and was carried here, perhaps on meteorites or interstellar dust, is known as the theory of panspermia. It seems unlikely that living organisms even remotely similar to the kinds we are familiar with would have survived the very radiation pressure that is presumed to have driven them to the earth. A modified version of this idea, directed panspermia, holds that intelligent beings outside our solar system purposely seeded the earth with primitive life forms that subsequently evolved into the various present-day organisms. Nevertheless, the possibility that terrestrial life was derived from extraterrestrial sources is thought to be slight and usually is set aside in discussions of the origin of life.

THE TASK

What is the actual transition from nonlife to life that scientists are trying to explain? The most primitive forms of cellular life on the earth today are the bacteria. Like *all* other life forms on our planet, these one-celled organisms possess a membrane at their periphery, store their hereditary information in deoxyribonucleic acid (DNA), and express that information as proteins via a ribonucleic acid (RNA) intermediate (Figure 6.1). The thousand or so different

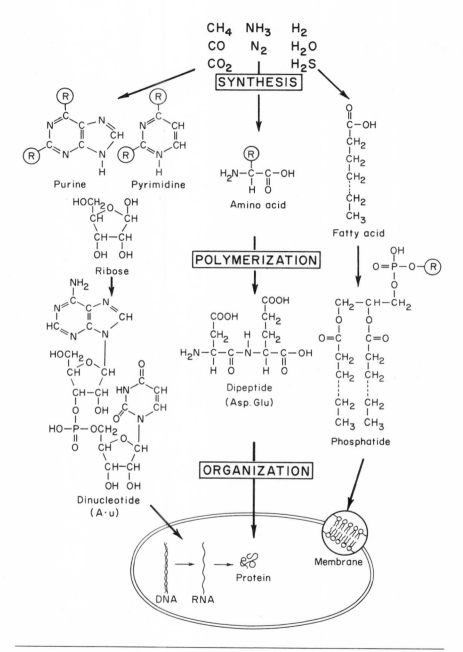

FIG. 6.1.   Three stages in the origin of life on earth.

proteins a bacterium can make direct the complex metabolic pathways by which the bacterium extracts energy from its environment and builds more copies of itself. Although the idea was entertained at one time, it is now considered highly unlikely that a chance assemblage of randomly synthesized prebiotic molecules could have been the source of the first bacteriumlike organism. The odds are overwhelmingly against it. Even these simplest of organisms are amazingly complex biological machines that must be immensely more sophisticated than the transitional forms that are thought to have bridged the gap between nonliving and living matter.

Even though present-day organisms bear little resemblance to their ancestors, they still provide our most reliable clues to the origin of life. That is to say, the genetic and metabolic processes common to the diverse modern organisms point the way to the basic mechanisms in the transition from nonlife to life. In all living cells, for example, the storehouse of genetic information, DNA, is composed of two intertwined chains of nucleotide building blocks. These nucleotides are composed of a purine or pyrimidine base linked to a phosphorylated five-carbon sugar, deoxyribose. The nucleotides of DNA contain only four bases: the two purines, adenine (A) and guanine (G); and the two pyrimidines, cytosine (C) and thymine (T). In double-stranded DNA, A always pairs with T, and G always pairs with C. These pairing rules, which were deduced by James Watson and Francis Crick in 1953, are the basis for accurate replication (duplication) of DNA. Information in DNA is stored in the linear sequence of nucleotides in the chain. In bacteria the genetic information is stored in one DNA molecule roughly a million nucleotides in length, whereas in humans it is stored in twenty-three pairs of chromosomes that are composed of nearly five billion nucleotides.

To be expressed, the genetic information in DNA must first be transcribed into another nucleic acid, RNA. RNA is similar to DNA except that uracil (U) takes the place of thymine and the five-carbon sugar, ribose, is used in place of deoxyribose. RNA molecules typically are single-stranded and are, at most, a few thousand nucleotides long. In general, individual RNA molecules contain the information for only one or a few of the thousands of informational units (genes) in the DNA.

The linear information in such RNA molecules is translated into proteins, which are chains of amino acid building blocks. The specific chemical properties of the amino acids in the chain force the protein to fold into a specific three-dimensional shape, which is required for carrying out a specific cellular function. Most commonly these proteins perform specific catalytic functions that collectively transform nutrients from the environment into the various molecules essential for cellular metabolism. The coding relationship between nucleotides in RNA and amino acids in protein is virtually identical in all organisms and is termed the genetic code. It was deciphered less than twenty years ago.

In modern organisms, the genetic apparatus and metabolic machinery are housed in cells. The periphery of every cell is marked by a membrane that separates the cellular contents from the outside environment. Such a division is essential for life as we know it. Membranes are composed of phosphatide (phospholipid) building blocks, which are made up of a three-carbon compound, glycerol phosphate, to which are attached two fatty acid tails and one of a few varieties of head groups. Phosphatides typically are hydrophilic (water-liking) at their heads and hydrophobic (water-hating) at their tails. In membranes, phosphatides are arranged into a double layer or bilayer. The hydrophilic heads are on the surface in contact with the water environment on either side; the hydrophobic tails abut one another in the interior, thus forming a hydrophobic barrier to the free flow of water soluble compounds into and out of the cell. It is this barrier that permits the internal environment of the cell to differ markedly from the external environment.

These remarkable common features along with some others buttress the evolutionists' belief that all organisms on earth had a common distant ancestor. Furthermore, they provide useful theoretical insights into how life might have arisen. They indicate, for example, that self-replication requires both nucleic acids and proteins: proteins catalyze the replication of DNA, and DNA encodes the proteins. As will be discussed, self-replication is the key to biological evolution. Thus the coding relationship between nucleic acids and proteins probably developed at the same time as replication. Because the present coding relationship is between RNA and proteins, many scientists believe that RNA was the first genetic material and that DNA and modern cellular membranes were later evolutionary developments. This viewpoint simplifies the problem somewhat; but before there could be RNA and proteins, the nucleotides and amino acids from which they are made first had to be synthesized.

The origin of life on earth can be considered conveniently in three successive stages: the synthesis of organic building blocks from the gases in the atmosphere, their linkage into polymers (chains), and the organization of these polymers into a primitive life form. In the gradual transition from gases to organic compounds, to polymers, to organized systems, the constituent atoms become arranged into increasingly complex patterns. This striking increase in order is cited by many creationists as a violation of the second law of thermodynamics. Such an objection, if valid, would eliminate the postulated sequence of events from scientific consideration.

The second law of thermodynamics states that the only possible processes are those in which the total entropy (disorder) of the universe increases. Since evolutionary events are said to progress in the opposite direction, toward decreased entropy, this law might be taken to suggest that evolution is impossible. The law demands, however, that we evaluate not only the entropy of the process under consideration, but also the entropy of the surroundings, because the *total* entropy change is the critical quantity. Since thermodynamics is covered in detail in another chapter, a single example will suffice here.

When a water solution of individual phosphatide molecules is mechanically agitated, it *spontaneously* forms highly ordered bilayer structures that closely resemble cellular membranes. This impressive increase in the order of the phosphatide molecules is more than balanced by a corresponding decrease in the order of the water molecules, because in solution the hydrophobic tails of the phosphatide molecules are surrounded by a sheath of water molecules whose freedom of motion is severely restricted. When these same hydrophobic tails are buried in the interior of a membrane, they are not in contact with water and the formerly restricted water molecules have increased freedom. Thus the entropy of the overall system—phosphatide molecules *plus* water molecules—increases substantially upon membrane formation. The illusion of increasing order and the specter of natural law violation exist only if we focus too narrowly on one component of the system. Properly evaluated, the sequence of events postulated by evolutionists is completely in accord with the second law of thermodynamics, and therefore creationists' objections on this point are unfounded.

Synthesis

The atmosphere and hydrosphere (water layer) of the primitive earth are thought to have been produced by volcanic outgassing of the earth's interior as it was heated by the accretion of new material during formation and by the decay of entrapped radioactive elements. Eventually a mixture of gases including CH_4, CO_2, CO, N_2, NH_3, H_2O, H_2S, and H_2 was released into the primitive atmosphere. Because the actual composition of such a mixture is strongly dependent on the temperature at which outgassing occurs, the exact composition of the primitive atmosphere is uncertain. The mixture of gases produced by modern volcanoes gives little information about the early atmosphere because the modern gases are thought to be produced from melted crustal rocks and recirculated groundwater.

The most critical question about the earth's primitive atmosphere is whether it was reducing or oxidizing. This question is crucial because it is only in a reducing or neutral atmosphere that organic compounds can be synthesized and can accumulate. Organic syntheses that are impossible in our oxygen-laden atmosphere are plausible in a nonoxidizing atmosphere. Undoubtedly there were at most only trace amounts of free oxygen (O_2) initially. The present high oxygen content of our atmosphere derives primarily from photosynthetic processes in living organisms.

A reducing or neutral atmosphere at moderate temperature does not produce significant amounts of organic compounds without a source of energy. On the primitive earth there were several potential sources of energy for organic synthesis, including sunlight, electric discharges, geothermal energy, cosmic rays, radioactivity, and shock waves. The most important energy sources were probably electric discharges and the ultraviolet components of sunlight. Ultraviolet light would have been most important for organic synthesis in the upper atmosphere and electric discharges for organic synthesis nearer the surface of the earth. In both kinds of synthesis, some of the input energy is trapped in a useful form by the production of chemically reactive compounds like hydrogen cyanide (HCN). These are called "reactive intermediates" because they combine with themselves to form more complex organic molecules. If deposited in the hydrosphere, these organic molecules would have been protected from the destructive forces of the energy sources that created them.

In 1953 Stanley Miller set up the first experiment to simulate prebiotic conditions on the primitive earth. A mixture of ammonia (NH_3), methane (CH_4), and hydrogen (H_2) gases was added to an evacuated flask containing water. The water was boiled to circulate the gases past an electric spark, and the products were condensed into a separate tube, where the nonvolatile ones remained while the volatile products recirculated past the spark. This apparatus is a crude model of the primitive earth: the mixture of gases represents the atmosphere, the water represents the hydrosphere, and the alternate cycles of heating and cooling in the presence of the spark simulate the weather. This simple experiment produced quite surprising results—a relatively uncomplicated mixture of organic compounds, amino acids being prominent among them. The products are uncomplicated because only a few varieties of reactive intermediates (primarily hydrogen cyanide and aldehydes) were generated by the spark.

Several related experiments using different energy sources and different gas mixtures have produced similar results, with many, but not all, of the twenty amino acids found in modern proteins represented in the products—glycine, alanine, aspartic acid, glutamic acid, and valine being the most abundant. Some of the main amino acids in contemporary proteins, including lysine, arginine, and histine, were either undetectable or present in only minute amounts. The rarity of these amino acids suggests that primitive proteins may have been composed of fewer varieties of amino acids than are modern proteins. Perhaps the missing amino acids had to await the development of biological processes for their synthesis.

A striking confirmation of the general validity of these prebiotic experiments comes from the analysis of organic compounds contained in meteorites. On September 28, 1969, a carbonaceous (carbon-containing) meteorite fell near Murchison, Australia. Portions were collected that same day and handled carefully to prevent contamination. The amino acids and other organic compounds found in this meteorite and others are remarkably similar in kind and relative abundance to those formed in the various prebiotic syntheses. Thus, the experimentalists' observations concerning prebiotic synthesis of amino acids appear largely correct.

The prebiotic syntheses of the purine, pyrimidine, and sugar components of nucleic acids are somewhat more difficult to explain than the synthesis of amino acids. The synthesis of adenine has been examined in some detail because it is not only a constituent of nucleic acids but also is used in the form of adenosine triphosphate (ATP), which is the main cellular intermediate in the utilization of metabolic energy. Adenine is a simple pentamer (that is, a complex molecule composed of five identical units, or monomers) of the reactive intermediate, hydrogen cyanide (HCN), and is synthesized readily under prebiotic conditions. It seems likely that four molecules of HCN first combine to yield a tetramer, which then rearranges in the presence of light and adds one more molecule of HCN to form adenine. The simplicity of adenine synthesis and its expected higher abundance in the prebiotic environment may underlie the selection of ATP as the principal energy carrier in modern organisms.

The synthesis of guanine and the pyrimidines—cytosine, uracil, and thymine—have not been investigated as intensively as that of adenine. However, guanine can be synthesized under plausibly prebiotic conditions from a tetramer of HCN by a hydrolysis reaction involving cyanogen (C_2N_2), which is another reactive intermediate produced by spark discharges. It is interesting that both adenine and guanine have been detected among the organic constituents of meteorites. The proposed prebiotic syntheses of cytosine, uracil, and thymine are rather less convincing. However, a substantial amount of uracil is produced from cytosine by hydrolysis in water, so if cytosine were synthesized prebiotically, uracil would have been present as well.

The synthesis of sugars occurs readily from formaldehyde (H_2CO), which is a common reactive intermediate formed under prebiotic conditions. The heterogeneous collection of sugars formed in this way does contain a small proportion of ribose, which is a simple pentamer of formaldehyde molecules. An unexpected difficulty arises in the attachment of ribose to a purine or pyrimidine base to form a nucleoside. No really satisfactory method for the prebiotic synthesis of nucleosides has been reported. Direct heating of ribose with bases either in the dry state or in aqueous solution yields little or no nucleosides. However, when purines are heated with ribose in the presence of certain inorganic salts, nucleosides are produced in reasonable amounts. One of the most efficient salt catalysts is obtained by evaporating seawater—a tantalizing observation. Nevertheless, the formation of nucleosides and nucleotides (a nucleoside to which a phosphate has been added) represents a major obstacle in prebiotic synthesis. Perhaps nucleosides and nucleotides were produced directly rather than by combination of free bases and ribose. These difficulties remain to be worked out.

Fatty acids, which are the main constituent of the phosphatides in biological membranes, also have proved to be difficult to synthesize under plausibly prebiotic conditions. The action of electric discharge on mixtures of methane and water produces high yields of two- and three-carbon acids but only minute amounts of twelve- to twenty-carbon fatty acids. In addition, these longer fatty acids are highly branched, unlike the straight-chain fatty acids in contemporary membranes. The difficulty in synthesizing fatty acids is an additional reason for suggesting that contemporary membranes were a later evolutionary addition. Perhaps the first life forms made use of branched fatty acids or suitable alternative molecules in their membranes until they evolved sufficiently to synthesize the more usual phosphatides.

To demonstrate a plausible prebiotic synthesis of important organic molecules is only part of the problem. The essential organic molecules would have had to accumulate to reach significant concentrations. The concentration of organic molecules in the primitive oceans would have depended on the balance between their rates of synthesis and their rates of decomposition. Because the rates of synthesis were limited by the energy sources available, only organic molecules with relatively long lifetimes would have accumulated. The more abundant amino acids produced in prebiotic experiments all have reasonably long half-lives (the time it takes for half the molecules to decompose), which range from tens of thousands of years to billions of years. The same is true for purines, pyrimidines, and fatty acids. The instability of ribose poses a serious problem, since it would have been hydrolyzed relatively quickly in the prebiotic oceans. However, its stability increases greatly when it is attached to a purine or pyrimidine base. Thus, if a prebiotic synthesis of nucleosides or nucleotides could be demonstrated, the instability of free ribose would present no real difficulty.

Despite the qualifying statements and fragmentary information, our current knowledge of prebiotic chemistry really is quite impressive. Nevertheless, evolutionists and creationists alike are well aware of the several difficulties I have pointed out and of some others I have glossed over because of space limitations. Evolutionists generally are optimistic and view any remaining difficulties as essentially chemical problems that are likely to yield to further experimentation. Creationists, on the other hand, regard these difficulties as intractable problems that constitute grave weaknesses in evolutionary theory. Much the same difference of opinion exists regarding the chemistry of polymerization, which is discussed in the next section.

Many creationists argue that these difficulties demonstrate the impossibility of an evolutionary origin of life and conclude therefrom that life must have been created. Such statements can be rebutted on logical grounds. First, difficulty cannot be equated with impossibility. Second, even if incontrovertible evidence against an evolutionary origin of life on earth were discovered tomorrow, this would not mean the only other alternative is to believe that life originated by supernatural creation. Indeed, other alternatives—a nonevolutionary natural origin (directed panspermia, for instance) or an unnatural origin—would still be possible. Obviously, evidence against an evolutionary origin would not establish the truth of any of these alternatives. Thus evidence against an evolutionary origin is *not* equivalent to evidence for a supernatural origin. Each explanation must stand up to scientific scrutiny on its own merits.

POLYMERIZATION

The joining of amino acids or nucleotides to form chains (or the construction of phosphatides, for that matter) requires the elimination of a molecule of water at each site of a new bond. Chemists refer to such reactions as condensations to suggest the removal of water. In a water environment, polymerization reactions require added energy and, therefore, are not favored. Modern organisms overcome this energy barrier by first activating one of the molecules to a higher energy state so that the actual joining reaction is favored. For example, in the construction of nucleic acid chains each new monomer unit to be added is activated to a high-energy triphosphate form. Upon joining the chain, two of the unit's phosphate groups are split off, thereby releasing sufficient energy to drive the reaction to completion. A similar strategy may have been operative in prebiotic polymerization reactions.

Several potential activating agents that would promote such condensation reactions have been produced in electric discharge experiments. These agents include cyanoacetylene, cyanogen, cyanamide, and the tetramer of hydrogen cyanide. Each has been used experimentally to promote condensation reactions, but with limited success owing to the presence of water. At each stage in such reactions, water competes directly with the condensation, thereby using up the condensing reagent or the activated intermediate without producing any significant polymerization. In general, for such reactions to occur in an aqueous environment, high concentrations of activating reagent and building blocks are required. Were the requisite high concentrations possible under prebiotic conditions?

The total quantity of potential organic material on primitive earth must have been enormous. If all the carbon now present on the surface of the earth, including the carbon present in carbonate rocks, coal, and oil, had been distributed as organic compounds in prebiotic oceans and lakes, it would have made an organic solution as concentrated as strong bouillon. However, this upper limit probably was never approached, and the concentrations of the essential organic molecules we have been considering would have been relatively low under any circumstances. Some mechanism would have been necessary to allow these organic compounds to reach sufficient concentration for polymerization. Several possible mechanisms have been suggested, including evaporation, adsorption onto surfaces, and freezing. Although none of these potential mechanisms for promoting polymerization has been explored adequately through experimentation, certain clay surfaces exhibit remarkable activities. Montmorillonite clays will promote the polymerization of activated amino acids into short chains, and bentonite clays bind the biologically common stereoisomers (mirror images) of certain amino acids and sugars. The properties of these surfaces are currently under experimental investigation.

An alternative suggestion is that polymerization was achieved by the heating of dry mixtures. In the laboratory the polymerization of short oligonucleotide chains and longer amino acid chains has been accomplished under such conditions. S. W. Fox, in particular, has extensively explored the thermal polymerization of polypeptides. When a dry mixture of amino acids is used in such experiments, substantial yields of polypeptides are produced. Thermal polypeptides even display limited catalytic activity in some reactions. Nevertheless, the role of thermal polymerization in prebiotic chemistry is open to question because there were probably few environmental niches in which such polymerization could take place. As should be apparent to the reader, the mechanism of prebiotic polymerization is a gap in the evolutionists' account of the origin of life. This area in particular will require further experimental exploration.

Many creationists fall back on probability arguments to illustrate the general implausibility of the evolutionists' position on prebiotic polymerization. One such argument focuses on the ten enzymes involved in glycolysis, a primary metabolic pathway by which glucose is converted to pyruvic acid in almost all contemporary organisms. The random, undirected polymerization of these enzymes from a mixture of the twenty amino acids is calculated to occur with a rough probability of $10^{-1,000}$. Even with relatively fast rates of polymerization and a billion-year time scale, it is argued, the likelihood that even one copy of each of these enzymes would be spontaneously produced is infinitesimal. The overall likelihood is not much improved even if only one of the ten enzymes is considered, and, of course, it becomes preposterously small for the thousand or so different enzymes in a typical bacterium. In the face of such overwhelming odds, how can evolutionists continue to assert the plausibility of life arising from nonlife?

To be meaningful, probability estimates must be calculated for the particular pathway under consideration. By way of illustration, consider the probability that I will arrive safely in London on a journey from New York. If I travel by jet plane or luxury liner, the probability is nearly 1; however, if I choose to swim, the probability is nearly 0. Here the possible and impossible exist side by side and both are correct estimates for the particular journey they characterize. The reason evolutionists are not swayed by the creationists' probability estimates is that the calculations are not based on the sequence of events that evolutionists have proposed. No evolutionist seriously proposes that even a single contemporary enzyme was polymerized prebiotically. Thus the probability estimates advanced to date by creationists are simply irrelevant to the evolutionists' arguments.

If evolutionists are not proposing the prebiotic polymerization of contemporary proteins, what are they proposing? Evolutionists suggest that, under prebiotic conditions, a more or less random collection of small peptides were polymerized, each peptide consisting of a few to a few tens of amino acids. The thermodynamically more stable species would have dominated the population. Some of these peptides would be expected to have catalytic activities, but such activities would have been much less efficient and less specific than is characteristic of the highly refined enzymes in contemporary organisms. Such a collection of primitive catalysts is important because it could modify the kinds and relative abundances of the organic molecules in the hydrosphere and, thus in principle, could supplement the strict chemical syntheses described in the previous section. In particular, these primitive catalysts may have been responsible for stabilizing and replicating nucleic acids, which is a necessary first stage in the development of a self-replicating system, as outlined in the next section. Once a primitive self-replicating system emerged, the process of improvement through natural selection could begin.

ORGANIZATION

Nucleic acids are the only molecules known to produce copies of themselves. The basis for this behavior is their inherent ability to pair accurately by Watson-Crick rules (A with T or U; G with C). During replication, one strand of a nucleic acid acts as a template upon which a second complementary strand is constructed. This second strand, in turn, can act as a template for polymerization of a copy of the original strand. These "plus-minus" pairs of molecules are the key to replication. In all modern organisms the genetic material is DNA, which is composed of two complementary strands wound together in a double helix. DNA replication in these organisms is an efficient and accurate process involving dozens of proteins. During replication, the plus and minus strands of the parental duplex are separated and copied to form two (usually) identical daughter duplexes. Although DNA holds center stage now, many scientists think that RNA was the original genetic material. One reason is that nucleic acid information today is deciphered into protein at the level of RNA, and, thus, the simplest hypothesis is that it originated that way. This viewpoint has much to recommend it.

The theory was developed by Manfred Eigen and his colleagues, who explored it experimentally using a bacterial virus in which genetic information is carried on a single-stranded RNA molecule a few thousand nucleotides long. This RNA strand is called the plus strand. Once it is introduced into a bacterium, the viral plus strand directs the synthesis of a protein called replicase. Replicase first copies the plus strand into the complementary minus strand and then copies the minus strand into more plus strands for the next generation of viruses. In one series of experiments the plus strand was mixed with replicase in vitro and replicated in the presence of activated nucleotides. Eventually, after many cycles of replication and dilution (to simulate growth and environmental destruction), small RNA molecules only one or two hundred nucleotides long became the dominant species. Surprisingly, similar RNA molecules eventually arose even if no RNA polymer was present initially. Thus when the ability to replicate is the only property selected for, as it was in these experiments, an RNA species evolves that is the most efficient target sequence for replication by replicase. When the experimental conditions were altered, a new RNA species arose that was optimal for that experimental environment. These simple RNA molecules arise because replicase makes occasional mistakes (mutations), and RNA molecules with mutations that favorably affect their rate of replication contribute more than their share to the next generation.

In the case of prebiotic times, the question is whether the primitive environment could have provided the necessary catalysts. In experiments designed to explore this question, Leslie Orgel and his colleagues have shown that a long chain of Cs in the presence of lead ions and a mixture of activated A and G nucleotides correctly polymerized G nucleotides more than 90 percent of the time. These results verify both the catalytic nature of the ions and the importance of Watson-Crick pairing rules under these conditions. When zinc ions were present, the copying process was much more accurate, and the G chains were up to forty nucleotides long. These simple ions undoubtedly were present in the primitive earth environment and could have catalyzed an inefficient replication of RNA polymers. Is it a coincidence that a zinc ion is present in all contemporary RNA polymerases (enzymes that catalyze polymerization of RNA)?

The strength of G-C pairing and the fidelity of polyC-polyG replication in the presence of zinc ions suggest that RNA polymers rich in G-C would have predominated in the early environment. The inaccuracy of any primitive replication system would mean that the dominant RNA varieties actually would be a collection of closely related sequences. Eigen and his colleagues have calculated that an RNA species longer than 50 to 100 nucleotides could not have maintained its information content against the mutation pressure of a mistake-ridden primitive replication process. Thus the early RNA polymers were most likely short sequences rich in G-C.

If the environment provided peptides that stabilized certain RNA polymers or catalyzed their replication more efficiently than zinc ions alone, those

RNA polymers would come to dominate the population. Such a population of RNA polymers would be expected to change their average sequence in response to changing environmental conditions on the primitive earth, much as they do in the laboratory. I term this arrangement *directed replication* to suggest the system's continuing dependence on the random polymerization of functional peptides by the environment. It seems unlikely that this random environmental process could be improved in a selectable fashion, and this limitation severely restricts the whole system's evolutionary horizons.

Over geologic time, directed replication would have generated a great variety of RNA sequences, each with a slightly different surface character. A momentous step in the origin of life was the eventual generation of an RNA surface that promoted the formation of a peptide with RNA replicator activity. Such an RNA polymer and its encoded replicator peptide constitute a system for *self-replication*, which would depend on the environment only for a continuing supply of precursor molecules. The principal advantage of such a system is that both the protein and nucleic acid components can be improved through mutation and natural selection. Improvements in the protein ultimately lead to more accurate replication of the nucleic acid component, which increases the potential for information storage. Therefore, these RNA polymers would qualify as true gene precursors, or *protogenes*.

If a self-replicating system is to improve by natural selection, it must be separated from other RNA molecules in the environment. Although other ways can be imagined, a natural way to achieve such a compartmentalization would be by enclosure within a primitive membrane, as might happen spontaneously in a greasy sea sufficiently agitated by wave action. Given a supply of precursors, the compartmentalization of a coding RNA would lead to a rapid, geometric replication as the concentrations of the coding RNA and the encoded protein both increased. By contrast, the compartmentalization of a noncoding RNA would lead to a slower linear replication rate, dependent on a fixed supply of replicator molecules. When these compartments released their contents back into the bulk environment, there would be an enriched supply of replicator molecules for the coding RNA. In a similar way, an occasional mutant RNA that encoded a more efficient replicator or possessed a target sequence more effective for replication would have a selective advantage over its parent RNA. Continual cycles of compartmentalization, replication, and remixing would permit the relatively rapid evolution of a self-replicating system. The same evolution does not occur in a homogeneous environment; compartments are the quanta (individual units) of natural selection. Note how closely these compartments resemble modern cells.

A compartmentalized self-replicating system could add new functions. For example, a replicator protogene together with a protogene encoding a catalyst of peptide-bond formation would replicate much more quickly than either would alone. These kinds of *coupled replication* systems could continue to "adopt" new protogenes that encoded useful catalytic functions, such as those for increased efficiency of replication or protein synthesis. In a modern bacterium, more than 100 genes — about 10 percent of its informational capacity — are devoted to replication and protein synthesis. Alternatively, a new protogene might be adopted by the dominant collection if it encoded a catalyst that increased the concentration of some limiting component for polymerization. Additions of this kind would mark the primitive beginning of intermediary metabolism.

If life actually originated in this way, there would have been other important milestones along the evolutionary path to contemporary life forms. At some point there must have been a transition from the storage of information in RNA to its storage in DNA. A related step was the linkage of separate protogenes into a single molecule, perhaps RNA initially, but eventually into DNA. These steps may have awaited the adoption of a protogene encoding a protein for converting an environmental precursor into deoxyribose for DNA. The copying of RNA into DNA is a well-understood process today. Indeed, it was discovered in certain animal viruses, whose single-stranded RNA genomes encode a protein for copying the RNA into double-stranded DNA.

A second important transition was the conversion from surface-coded alignment of amino acids to the adapter system that cells use today. Contemporary cells correctly align the amino acids in proteins, not by direct use of an RNA surface, but through a set of adapter molecules called transfer RNAs (tRNAs). Each distinct tRNA molecule is linked at one end to a particular amino acid by enzymes that identify the correct tRNA for linkage by virtue of its surface features. At its other end, each tRNA recognizes specific sequences three nucleotides long, called codons, in an information-bearing RNA (messenger RNA) through nucleic acid-base pairing. In this way the sequence of codons in the message establishes the sequence of amino acids in the protein. Contemporary tRNA molecules are rich in G-C and average about eighty nucleotides in length, much like the protogenes we have been discussing. The conversion of protogenes into adapter molecules seems like such a natural evolutionary step that scientists have examined the possibility at some length. Computer analysis of the nearly 200 known tRNA sequences has revealed a distinct pattern of purines, mostly Gs, repeating at three nucleotide intervals. Significantly, in the present genetic code the codons beginning with G stand for glycine, alanine, valine, aspartic acid, and glutamic acid. Remember this list? These are the most abundant amino acids in meteorites and prebiotic synthetic reactions.

A third important transition was the acquisition of a stable cellular membrane. As discussed earlier, this step may have awaited development of a primitive phosphatide metabolism. Another problem is how to account for the evolution of contemporary membranes, which are not simple barriers but quite complex structures that permit nutrients to enter and waste products to leave. The selectively permeable character of modern cellular membranes is conferred by a variety of transport proteins that form molecule-specific channels through the hydrophobic interior of the membrane. A workable exchange of material with the environment would have had to develop before a truly stable membrane could impart a selective advantage to the system.

This series of transitions would mark the emergence of an essentially modern cellular life form. Although certainly a speculative sequence of events, it is characterized by small incremental steps that individually appear manageable. The critical transition in this proposed sequence, which is the development of a surface-coding relationship between an RNA and a replicator peptide, has not yet been accomplished in the laboratory. However, it seems inescapable that this or some analogous coding relationship is essential for the development of a self-replicating system that can be compartmentalized and improved through the process of natural selection. Once natural selection begins to operate on a primitive self-replicating system, evolution becomes a race for survival in which the reproductively fittest members contribute most to the subsequent generations.

This chapter began with a rather broad definition of science as the observation, experimental investigation, and theoretical explanation of all phenomena — natural, unnatural, or supernatural. Philosophers might argue instead that only natural phenomena are in the province of science. The problem is that in the entire history of science few, if any, unnatural or supernatural phenomena have been documented. Thus, it is not surprising that unnatural and supernatural phenomena are not represented in the realm of current science, but that fact does not eliminate them from the province of science. These statements are equivalent to saying that everything we understand about nature obeys natural laws that have never been observed to vary. The time scale of observation of these laws is not as restricted as it may seem, because when we look out into the cosmos, we are looking far back in time. To the extent such observations permit, there is no evidence that these laws have ever been different than they are today.

The single assumption that these natural laws have always been true leads to a surprisingly unified picture of the universe, which holds that matter has gradually changed from a high-temperature, high-density state ten to twenty billion years ago to its present low-temperature, low-density state. We call this change evolution, and nearly all scientists regard it as fact. Evolution is one of the strongest principles in science and is supported by independent and self-consistent evidence, both observational and theoretical, from astronomy, geology, and biology. That is not to say we understand the precise mechanisms by which evolution produced the observable features of the universe. These mechanistic questions are argued vigorously in many disciplines of science. Although this type of debate may look chaotic from the outside, it is part of the healthy refinement process characterisitc of science.

In this chapter we have examined one particular evolutionary transition, namely, the origin of life from nonliving matter. Evolutionists think the mechanism involved three stages. The first two, synthesis and polymerization, were chemical processes powered by the available energy sources. The third, organization, was a biological process initiated by the development of a compartmentalized self-replicating system. Although the mechanistic details are not yet complete, the proposed sequence of events is plausible and consistent with the laws of nature. As we have seen, the creationist objections that an evolutionary origin of life violates the second law of thermodynamics and the laws of probability can be rebutted on scientific grounds. On the other hand, creationist objections to the lack of completeness of mechanistic details raise an important question about the role of observation in science.

Although observations are critical, the real currency in science is explanation, because explanations tie many observations together into a tentative understanding. Particularly powerful scientific explanations gain the rank of theory. Strong theories in science are not displaced by discrepant observations but only by stronger theories that include the discrepant observations and thereby improve understanding. For example, certain eccentricities in the orbit of Mercury that are inconsistent with Newton's theory of gravity had no influence on that theory until Einstein incorporated them into a more powerful explanation of gravity. Creationists have cited particular observations, which they consider inconsistent with, for example, an evolutionary origin of life; to replace evolutionary theory, they propose that the universe must have been put in place by a supernatural creator in a fully functional form much as we observe it today. The trouble with the creationists' proposal is that it generates many more observational discrepancies than it rectifies. Furthermore, the discrepancies are of such magnitude that powerful scientific explanations in astronomy, geology, and biology would have to be discarded. Our scientific understanding of nature is simply not compatible with the suggestion that the universe began much as it is now.

Because more explanatory power would be lost than gained, science judges the creationist proposal to be weak and discriminates against it. Although the word "discrimination" has an unfair ring to it because it raises the specter of social discriminations that are abhorrent to us all, discrimination among theories is an essential component of the scientific process; an even-handed treatment of weak and strong theories alike would create a terrible muddle. Such a balanced treatment would be contrary to the goal of scientific investigation, which seeks, not to list all possible explanations, but rather to discover the *best* explanations for the phenomena of nature. Creationism in its popular form is a weak scientific explanation. For this reason alone, creationism does not attract widespread scientific attention.

FOR FURTHER READING

BERNAL, J. D. *The Origin of Life*. New York: World, 1967.
Offers a physicist's view of life's origins with especially good historical and philosophical discussions and includes the original papers by Haldane and Oparin as appendixes.

DICKERSON, RICHARD E. "Chemical Evolution and the Origin of Life." *Scientific American* 239(3):70–86.
Presents a lucid description of prebiotic chemistry with simple diagrams to illustrate clearly the transition from gases to organic molecules.

EIGEN, M.; GARDINER, W.; SCHUSTER, P.; AND WINKLER-OSWATITSCH, R. "The Origin of Genetic Information." *Scientific American* 244(4):88–118.
Provides a theoretical and experimental framework for understanding the biological side of the origin of life, namely, the origin of the coding relationship between nucleic acids and proteins.

GISH, DUANE T. *Speculations and Experiments Related to Theories on the Origin of Life (A Critique).* San Diego: Institute for Creation Research, 1972.
  Presents the definitive account of creationists' objections to the experimental and philosophical underpinnings of evolutionary science as it relates to life's origins.
MILLER, STANELY L., AND ORGEL, LESLIE E. *The Origins of Life on Earth.* Englewood Cliffs, N.J.: Prentice-Hall, 1974.
  Analyzes carefully the prebiotic plausibility of chemical syntheses carried out in the laboratory and pinpoints the problem areas that must be dealt with to provide a complete account of the origin of life.

[7]

The Evolution of Life

ROBERT H. CHAPMAN

A key question in the creation-evolution controversy concerns what has happened since life originated on the planet. The creationists maintain that the initial living products of creation were structurally complete "kinds," sometimes referred to as "baramins."[1] These kinds of living things were all created at the time of creation and have remained essentially unchanged in the years since. The creationists acknowledge that variation may exist within a kind, but one kind of organism, plant or animal, cannot change into or merge with another kind; nor have any new kinds arisen since the creation. Some kinds have become extinct since their creation, with many of these extinctions occurring at the time of a worldwide flood as described in Genesis.[2]

We see in modern plants and animals some degree of adaptability, which refers to the ability of populations of organisms to adjust to changes in their environment. The adjustment may take the form of increases or decreases in the relative frequencies of morphological or physiological characteristics in the population. Creationists maintain that the adaptability of modern organisms results from genetic variability that was included in life forms at the time of creation by the creator. Possible changes in organisms due to this genetic variability are still limited only to changes within a kind.

Creationists give no clear-cut definition of what constitutes a kind. The example cited most often in the creationist literature is that of the dog kind, the original representative of which was perhaps an organism much like the modern wolf. From this organism came the modern wolves, coyotes, jackals, and all the breeds of domesticated dogs. The important point, according to the creationists, is that all this diversity is still limited to the dog kind. Human beings are placed as a separate kind. Other kinds are generally held to be intuitively obvious.

These creationist tenets lead to two specific predictions about the history of life on earth. One is that at a particular point—at the time of creation and for some period after that—all possible kinds of organisms were living on the face of the earth. They coexisted, not in the sense that they all necessarily lived in the same area but in the sense that during this time period all kinds of plants, animals, fungi, and microbes were in existence, including the human race.

A second prediction following from the creationist tenets is that the number of kinds in existence should have declined continuously from the time of creation until now. This derives directly from the belief that no new kinds can be formed but that some kinds—for example, the dinosaur kind—can become extinct. Also, because one kind cannot change into another kind, no organisms in the present or past can be intermediate between two kinds.

These predictions can be tested by examining the fossil record, which provides us with evidence about the physical appearance of different organisms and the approximate time during which these organisms existed. (See Chapter 5 for more information on the formation of fossils.) This evidence readily falsifies the predictions derived from creationist tenets.

Figure 7.1 represents a summary of the evidence revealed by the fossil record. The eras, periods, and epochs are somewhat arbitrary time periods, delineated and named by paleontologists for convenience in referring to events in the past.

Consider first the creationist idea that early in the history of life on earth all the different kinds coexisted. Figure 7.1 shows that this was not the case. The earliest forms of life were simple, single-celled aquatic organisms, and no living organism existed on land surfaces. Over time more complex organisms came into existence and colonized the land. We can identify the approximate time when groups such as reptiles and mammals emerged. Note that such dominant organisms as humans and flowering plants are relative newcomers. The important point is that all these kinds were never in existence at the same time. For example, we find no evidence in the abundant fossils of dinosaurs and humans that the two existed during the same time periods. Some of the creationist literature claims that human footprints have been found in ancient trilobite beds and that human and dinosaur footprints crossing paths have been found preserved in the same rock.[3] Paleontologist Robert Sloan has summarized the evaluation of these footprints by the scientific community:

> If there were any further truth to the association of true human or humanoid footprints with pre-Pleistocene fossils, it would make the scientific reputation of the paleontologist who could demonstrate it. But the footprint would have to be demonstrably impressed in the original wet sediment with appropriate sedimentary deformational features, would have to be anatomically accurate, preferably excavated as part of a continuing trackway to show strides, and preferably continuing unaltered and undisturbed in the sediments of the next younger bed of the same geologic age. None of these criteria are fulfilled by the examples cited in the Creationist literature. They all appear to be hoaxes.[4]

Even if this creationist claim were true, the fossil record would still unambiguously demonstrate that all organic kinds have never coexisted.

The idea that the number of kinds decreased over time is also false according to Figure 7.1. The number of kinds increased slowly and steadily until the

ERA	PERIOD	EPOCH	EVENTS
QUATERNARY	PLEISTOCENE	EVOLUTION OF MAN
TERTIARY	PLIOCENE MIOCENE OLIGOCENE EOCENE PALEOCENE	MAMMALIAN RADIATION
CRETACEOUS		LAST DINOSAURS FIRST PRIMATES FIRST FLOWERING PLANTS
JURASSIC		DINOSAURS FIRST BIRDS
TRIASSIC		FIRST MAMMALS THERAPSIDS DOMINANT
PERMIAN		MAJOR MARINE EXTINCTION PELYCOSAURS DOMINANT
PENNSYLVANIAN		FIRST REPTILES
MISSISSIPPIAN		SCALE TREES, SEED FERNS
DEVONIAN		FIRST AMPHIBIANS JAWED FISHES DIVERSIFY
SILURIAN		FIRST VASCULAR LAND PLANTS
ORDOVICIAN		BURST OF DIVERSIFICATION IN METAZOAN FAMILIES
CAMBRIAN		FIRST FISH FIRST CHORDATES
EDIACARAN		FIRST SKELETAL ELEMENTS ● FIRST SOFT-BODIED METAZOANS FIRST ANIMAL TRACES (COELOMATES)

ERAS (left axis): CENOZOIC, MESOZOIC, PALEOZOIC, PRECAMBRIAN

CARBONIFEROUS (spanning PENNSYLVANIAN and MISSISSIPPIAN)

MILLIONS OF YEARS AGO: 0, 50, 100, 150, 200, 250, 300, 350, 400, 450, 500, 550, 600, 650, 700

FIG. 7.1. The divisions of the geologic timetable are given here along with some of the major evolutionary events that happened during these periods of time. Used by permission from J. W. Valentine, "The Evolution of Multicellular Plants and Animals," *Scientific American* 239(September 1978):142.

Cambrian period, at which point in the geologic record we see evidence of a relatively rapid period of adaptive radiation that produced many new kinds of organisms. Since that time many life forms have become extinct and many new kinds—such as birds, reptiles, mammals, humans, flowering plants, and so on—have developed.

Creationists claim that the arrangement of fossils in sedimentary rocks is the result of a sorting process in the Noachian Flood.[5] This curious hypothesis suggests that almost all sedimentary rock formations were laid down as the floodwaters described in Genesis receded. All fossils were then deposited in the resulting sediment. Creationists suggest that the observed ordering of the fossils is due to the ability of organisms to avoid the rising floodwaters or to survive in the water. The fossils found in the higher sedimentary layers were either more complex and thus able to resist the floodwaters or to avoid them for longer periods of time, or were living at higher altitudes to begin with.

Studies of the fossil record show that this idea is clearly untenable. Aside from the fact that various dating techniques show that the sedimentary rock formations could not have been formed at the same time, the sorting hypothesis could not account for the observed formations, as is obvious from Figure 7.1. Why would dinosaurs, for instance, end up higher than fish? Why would flowering plants end up higher than ferns? Why would flying dinosaurs like the pterodactyls end up lower than the giant sloth? The turbulence of the receding floodwaters would, of course, cause some mixing of the organisms. But the above arrangements are consistent in the fossil record. This consistency would not occur with sedimentation from a single large flood. Even with the sorting process, we would expect to find occasional examples of humans and dinosaurs in the same stratum, flowering plants and trilobites together, and some mammals, reptiles, or fish in the lowermost strata. This is never the case.

The idea that different kinds were created as different and remain different—that is, they never combine with each other or form new kinds—implies that present-day organisms should exhibit some degree of genetic discontinuity. This means that there should be distinct genetic differences from kind to kind.

The genetic relationship between different organisms can be determined to some extent by comparing the amino acid "sequences" of proteins in one organism with the sequences in others. Because these amino acid sequences are dictated by the genetic code, any differences in sequence or in length of proteins are most likely due to genetic differences.

The protein molecule that has been studied most extensively is cytochrome C. Cytochrome C is a relatively short, easily extracted, "electron transport" molecule involved in respiration. This molecule is found in various organisms, including fungi, animals, and plants. A comparison of the cytochrome C molecules from different organisms indicates that critical sites on the molecule never vary, whereas other sites on the molecule vary greatly from organism to organism. These variations, or amino acid "substitutions," result from mutations in the nucleic acid sequence coding for cytochrome C and do not affect the function of the molecule.

When the cytochrome C molecules from a large variety of organisms are compared, it can readily be seen that organisms that we intuitively regard as being similar to each other have fewer differences in their cytochrome C molecules. Again, these differences are not functional. If all of these organisms at one time had a common ancestor, the entire pattern makes sense. We can see that the accumulation of differences between two types of organisms over time is proportional to the time since those organisms have diverged from each other.

On the basis of the number of amino acid differences in the cytochrome C molecule, this "molecular clock" can then be used to construct a phylogenetic diagram, which shows a pattern of evolution (Figure 7.2). For finer distinctions within a group, rapidly evolving proteins such as fibrinopeptides may be used.

Some creationists argue that these genetic similarities reflect the fact that the creator simply used similar building materials for similar types of organisms. In this case, however, the genetic differences are not functional. The accumulation of differences is due only to the background mutation rate and does not help or hurt the organism. It seems unlikely that a creator would create this kind of pattern. Other chemical comparisons as well as comparative morphology and physiology show the same hierarchical pattern of genetic resemblances.

The sequence of fossil organisms over time and the observed patterns of genetic relationships have been explained by a collection of principles referred to as the "theory of organic evolution." Theory in this case simply refers to this collection of ideas. The ideas embodying organic evolution are diverse and complex. Scientists agree almost totally about some aspects of the theory, but strongly disagree about others. There is relatively little disagreement about whether organic evolution has actually occurred, that is, whether modern organisms have descended from a common ancestor. Most scientists consider the evidence for this to be overwhelming.

Scientists define organic evolution as a change in the genetic basis of populations of organisms over time. It has been known since 1900 that the frequencies of alleles (forms of a gene) and genotypes (combinations of alleles) remain constant from generation to generation unless disturbed by one or more possible factors. This genetic stability is usually called the Hardy-Weinberg equilibrium. The allele and genotype frequencies are mathematically related to each other if the population is in Hardy-Weinberg equilibrium. This relationship is very simple mathematically and can be described in terms of the expansion of a polynomial.

Any factor or combination of factors that causes a population to deviate from Hardy-Weinberg equilibrium is, by definition, a mechanism of organic evolution. These factors are easy to list. The questions about evolution that occupy most researchers, and therefore determine the nature of their experiments, concern the relative importance of the different factors in different situations.

One of these factors, natural selection, was first proposed by Charles Darwin and Alfred Russel Wallace at a session of the Linnean Society of London on July 1, 1858. Darwin and Wallace had arrived at the idea independently after each had spent years studying populations of plants and animals in their natural habitats. A year later, Darwin published his famous book *The Origin of Species,* in which he elaborated his ideas and presented many of his observa-

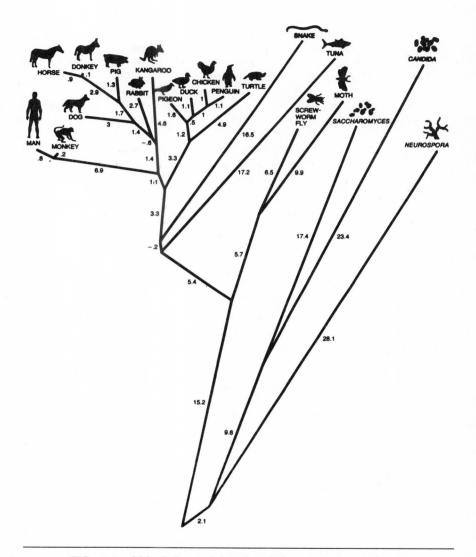

FIG. 7.2. This phylogeny is based on differences in the amino acid sequence of cytochrome C from each species in the diagram. The numbers on the branches are the minimum number of nucleotide substitutions in the DNA of the genes that could have given rise to observed differences in amino acid sequence. Used by permission from F. J. Ayala, "The Mechanisms of Evolution," *Scientific American* 239(September 1978):69.

tions. Darwin had noticed that not all the individuals of a species were exactly alike. Any characteristic, he reasoned, will become more prevalent in the population if the individual possessing it produces a larger progeny that survives to adulthood than individuals not having the trait. This is the process that Darwin called natural selection.

The ability of an organism to leave more surviving progeny is termed evolutionary fitness. There are a number of different ways in which fitness may be increased. A particular combination of genes might increase an organism's ability to survive. The increased survivorship would thus allow the organism to produce and, in some cases, to protect more offspring. Alternatively, a genotype might increase the number of fertilized eggs or seeds produced per generation without affecting survivorship. The key to evolutionary fitness, therefore, is the ability of an organism to make a genetic contribution to the next generation. Evolutionary fitness may have nothing to do with physical strength or other characteristics that we usually associate with the term fitness.

The concept of evolutionary or Darwinian fitness has occasionally been criticized as being a tautology. A tautology is an axiom whose conclusion is implicit in its premises. In this case, if evolutionary fitness is defined as "those that survive" and evolution is described as the "survival of the fittest," then what is actually being said is that evolution is "the survival of those that survive," which is a tautology. Fitness, however, can be estimated not just after natural selection has operated, but before. Genetically similar organisms can be studied and demographic tables can be constructed that give average measures of such factors as number of viable offspring and the ability to survive in certain environments. With this information, evolutionary fitness can to some extent be predicted, and thus the tautology disappears.

A characteristic that increases the ability of an organism to cope with its environment is called an adaptation. Frequently, but not always, natural selection increases the adaptiveness of a population. Adaptive characteristics might be morphological, such as the thick fur of polar bears. Or they might be physiological, such as the requirement of an extended cold period for the germination of some seeds, which prevents germination during a brief period of warmth in the fall; a plant that germinated at this time would die in the cold of winter.

For natural selection to operate, genetic variation must be present. Many, but not all, types of natural selection will reduce the genetic variation in natural populations. For this reason, the generation of variation must be a continuous process. The ultimate source of genetic variation is mutation. A mutation is a change in a gene or chromosome. Some mutations are extremely rare and usually have no evolutionary significance because they are lost from the population. Other mutations occur with a measurable frequency. Many of these mutations result in the formation of a new allele. The accumulation of these alleles over a period of time is the ultimate source of genetic variability.

Mutations are random in the sense that there is nothing controlling their occurrence or direction. As one would expect, most changes in an organism's genetic complement are harmful. These harmful mutations may be slightly deleterious or they may be lethal. Their frequency in the population depends on several factors, such as how frequently the mutations occur, and on whether their action is compensated for by the presence of a normal allele. The frequency also depends on the effect of natural selection against the mutation, whether natural selection acts strongly against it or only weakly. As one would also expect, mutation occasionally produces an allele that is an improvement over the previous product. These mutations will increase in frequency as a result of natural selection.

Although the ultimate source of all change in genetic information is mutation, new combinations of genes (genotypes) are continually arising in sexually reproducing populations by the process of recombination. This process is the reshuffling of the genes of the parents in the offspring. Recombination is important for two main reasons. First, different genes of an organism interact with each other, and certain combinations of genes may give an organism a higher evolutionary fitness than other combinations. Second, the number of possible genotypes produced by recombination is nearly infinite. Recombination, therefore, can produce more different genetic types and at a faster rate than mutation can.

The study of the evolution of protein molecules has shown that the substitution of some amino acids for others in the protein chain has no effect on the function of the protein. This type of mutation is called a neutral mutation because it is neither beneficial nor harmful to the organism and thus it is not acted on by natural selection. The presence of some mutations of this type is indisputable. An example is found in cytochrome C. Evolutionists are currently debating the question of what proportion of the observed molecular variation is neutral and what proportion is under the influence of natural selection. An important point is that even scientists who argue that a large proportion of the observed molecular variation is neutral do not downgrade the role of natural selection in the adaptive evolution of organisms.

In a relatively small population of organisms, frequencies of alleles participating in fertilization events may not reflect precisely the frequencies of alleles in the population as a whole. This chance deviation from expected frequencies is called genetic drift. Population size is the key factor in determining the significance of genetic drift. In large populations, the deviation from expected frequencies is extremely small and genetic drift is therefore insignificant. In small populations, genetic drift may be the single most important factor in determining the frequencies of some alleles.

Several other factors may be involved in evolutionary changes in populations. Mating patterns within populations may have a profound impact on the gene pool of the population. If, for example, a population is inbred, recessive harmful alleles may be exposed more often. In a highly outbred population, these alleles may not be expressed when paired with a normal allele. There may also be a significant movement of individuals, spores, or gametes into or out of a population. This phenomenon is called gene flow and will affect genetic differences or similarities between populations.

Speciation, the formation of new species, is a process of genetic differentiation. All the processes discussed above may play a role in speciation. Many attempts have been made over the years to define a species in a way that will reflect the degree of genetic differentiation observed in natural populations.

The best-known attempt has been to define a species as a group of actually or potentially interbreeding organisms. This definition is based on the assumption that reproductive isolation (the inability of two types of organisms to produce fertile offspring) reflects a certain degree of genetic differentiation and will lead to even further differentiation. This key assumption is the reason for difficulties with this definition. Reproductive isolation may result from very slight genetic differences, or it may not be present even though genetic differentiation may be extensive. Species limits are of necessity largely mental constructs imposed on many organisms for the convenience of the biologist. Once this is acknowledged, taxonomic decisions at the species level can be made on the basis of the judgment and experience of the taxonomist with regard to the range and nature of the genetic variation in a group of organisms.

The collective group of ideas discussed here has been called the "synthetic" theory of evolution. Recently a great deal of attention has been given to a new theory of macroevolution, which deals with major changes in organisms over geologic time. The new theory is called "punctuated equilibria." According to this theory, macroevolutionary changes are not mainly the result of the steady accumulation of slight differences (phyletic gradualism). Rather, long periods of relatively little change are punctuated by short periods of relatively great change (punctuated equilibria). The proponents of punctuated equilibria have equated phyletic gradualism with the synthetic theory of evolution. This is a mistake. The components of the synthetic theory of evolution as discussed above are compatible with both phyletic gradualism and punctuated equilibria.[6]

Creationists maintain that the synthetic theory of evolution inadequately accounts for what we know about life today. Many of these questions are dealt with in other chapters in this book, for example, those on human evolution, thermodynamics, and the origin of life. Their three main objections to the ideas dealt with in this chapter are: mutations are inadequate to account for the generation of useful genetic variation; there is no way for one kind to change into another kind; and there are no intermediate forms in the fossil record.

The first objection has been voiced by J. N. Moore:

> Because the vast majority of mutations are lethal or cause impairment of the physiology of the organism and because the gene-mutation hypothesis suffers from the difficulties of the pathologic nature of and the great rarity of mutational changes, it follows that mutations are not useful as supporting evidence for the general evolution model; that is, "molecules to man."[7]

This statement reveals a profound misunderstanding of the population dynamics involved. The following quotation from an evolution textbook shows that mutation theory easily accounts for the observed rates of evolution:

> While rates of mutation vary enormously from one gene locus to another, and can also be greatly influenced by the environment, a rate of one mutation per gene locus in every 100,000 sex cells is a conservative estimate. Because all higher organisms contain at least 10,000 gene loci, and most of them contain many more, we can conservatively say that one individual

out of ten carries a newly mutated gene at one of its loci. As already pointed out, the great majority of these mutations are deleterious, but a small proportion of them are beneficial. From various experimental studies we can arrive at a conservative estimate of the proportion of useful mutations as one in a thousand.

On the basis of these estimates we can calculate that in any species about one in ten thousand individuals in each generation would carry a new mutation of potential value in evolution. Using conservative values of 100 million as the total number of individuals per generation and 50,000 as the number of generations in the evolutionary life of the species, we could expect that at least 500 million USEFUL mutations would occur during this life span. We do not know how many new mutations are needed to transform one species into another, but five hundred is a reasonable estimate. On this basis, only one in a million of the useful mutations or one in a billion of all mutations which occur needs to be established in a species population in order to provide the genetic basis of observed rates of evolution.[8]

Another creationist objection to evolutionary theory is that there is a genetic stability to creationist kinds.[9] The basis of this belief seems to be a denial of the mutation theory and citation of so-called "living fossils." Examples of reproductive isolation are cited to show that these kinds are genetically different. As explained earlier, reproductive isolation is due simply to genetic differences in specific genes and is not proof that one kind cannot change into another. In fact, molecular biology has found no mechanism for limiting genetic change. If we look at the genetic code of living organisms, we see a remarkable uniformity in the underlying mechanisms. The difference between a cow and an oak tree, for example, is due simply to a difference in the sequence of nucleotides in the DNA. Mutations provide the method for changing the nucleotide sequence. There is nothing that would limit the possible change at the boundary of a particular kind. So-called living fossils such as the horseshore crab, the tuatara, and the coelacanth do not contradict evolutionary theory. Tbse organisms inhabit an environment that has undergone very little change. They have few predators and a generalized diet. They have been able to survive in limited numbers and have not been affected by any factors that would cause them to change.

Finally, many creationists claim that although evolutionary theory suggests that millions of intermediate forms should be found in the fossil record, "there is no evidence that there have ever been transitional forms between these basic kinds (plants and animals)."[10] It must be pointed out first that evolutionary change does not occur at a uniform rate but may proceed quite rapidly at some times and quite slowly at others. This pattern might be responsible for some of the gaps observed in the fossil record. Other reasons for the gaps might be that an organism lived in a habitat unfavorable to fossilization or that a particular layer of sediment has been eroded or is otherwise inaccessible to paleontologists. Therefore, evolutionary theory does not predict "millions of intermediate forms."

There are good examples, however, of transitional forms. Transitional forms are well known between all higher groups of fishes. The icthyostegids of the late Devonian (see Figure 7.1) are intermediate between fishes and amphibians. Work in the Carboniferous has revealed a transition between advanced amphibians and primitive reptiles. *Archeopteryx* is clearly part bird and part reptile. The transition from reptiles to mammals is known even more clearly; the distinction between the two classes is chosen arbitrarily as the point at which the jaw joint between quadrate and articular bones was replaced by that between squamosal and dentary elements.

In conclusion, then, an examination of both the fossil record and living organisms falsifies the hypotheses of creationists. It is clear that organisms have changed over time, and all the evidence points to a common ancestor. The mechanisms that have been proposed and that continue to be incorporated into the modern theory of evolution adequately account for the patterns observed in modern and fossil organisms. Extensive efforts by creationists to discredit the theory of evolution have failed to come up with anything of scientific substance.

Evolutionary interpretations are important in explaining observations in many fields of biology, ranging from physiology to behavior. As we learn more about the world around us, the insight of the famous biologist Theodosius Dobzhansky seems increasingly justified: "Nothing in biology makes sense except in the light of evolution."[11]

FOR FURTHER READING

AYALA, FRANCISO J. *Molecular Evolution*. Sunderland, Mass.: Sinauer Associates, 1976.
  A collection of papers by experts in this field.
AYALA, FRANCISCO J. "The Mechanisms of Evolution." *Scientific American* 239 (1978):56–69.
  A very readable article summarizing current evolutionary theory.
DOBZHANSKY, THEODOSIUS; AYALA, FRANCISCO J.; STEBBINS, G. LEDYARD; AND VALENTINE, JAMES W. *Evolution*. San Francisco: W. H. Freeman, 1977.
  A textbook written for college juniors and seniors with a good background in biology.
FITCH, WALTER M., AND MARGOLIASH, EMANUEL. "Construction of Phylogenetic Trees." *Science* 155(1967):279–84.
  An advanced article that demonstrates how phylogenies are generated from molecular data.
LEWONTIN, RICHARD C. *The Genetic Basis of Evolutionary Change*. New York: Columbia University Press, 1974.
  An excellent book of interest to both the layperson and the scientist.
LEWONTIN, RICHARD C. "Adaptation." *Scientific American* 239(1978):212–30.
  An article that addresses the many misconceptions about this topic.
MAYR, ERNST. "Evolution." *Scientific American* 239(1978):46–55.
  This article takes a historical approach to the topic.
RAUP, DAVID M., AND STANLEY, STEVEN M. *Principles of Paleontology*. San Francisco: W. H. Freeman, 1978.
  A textbook by two prominent paleontologists with a good discussion of the theoretical considerations involved.
VALENTINE, JAMES W. "The Evolution of Multicellular Plants and Animals." *Scientific American* 239(1978):140–58.
  A well-illustrated article summarizing the major evolutionary changes in these two groups.

[8]

The Origin and
Evolution of Humankind

JOHN BOWER

THIS chapter focuses on the history of the organisms for which we tend to have the greatest concern—namely, ourselves. It is fitting that we should consider closely the opposing views on human evolution, for this is surely the heart of the issue toward which this book is addressed. Would there be much fuss about evolution if the concept applied to worms, shrimp, snails, ferns, and porcupines, but not to humans?

If our interest in human evolution were limited to a more or less narcissistic preoccupation with ourselves, it would be trivial. But there is much more than vanity at stake, for our views concerning our own origins often play an important role in political thought. This has been conspicuously evident in some nations whose course of development has been sustained partly by propaganda based upon notions of "racial destiny." It is more subtly apparent in democratic countries, where much political debate can be traced to differing basic assumptions about "human nature" and how it got to be however it is. Although the topics of this chapter are usually debated in the contexts of science and education, it is important to realize that they also have political implications.

THE CREATIONIST PERSPECTIVE

No one will be surprised to learn that the topic of human origins is extensively discussed in the creationist literature. Despite its fairly common appearance, the subject often seems to be treated in staccato fashion, with here a burst on one fossil find and there a burst on another. One important exception occurs in Duane Gish's book *Evolution? The Fossils Say No!*, which contains a chapter entitled "The Origin of Man" that is a more or less comprehensive critique of evidence for human evolution in the fossil record.[1] The following description of the creationist perspective on human origins is based largely on the material in that chapter.

The creationists seem to have a great deal more to say about how humans did *not* originate than about how they did, presumably because they are convinced that everything that really needs to be said about human origins has been conveyed in the Bible. This point deserves special emphasis in view of recent attempts by creationists to disassociate their position from any particular religious orientation. All doubt about the Judeo-Christian source of the creationists' convictions is removed by Gish's chapter on human origins. In commenting on a 1973 lecture by Richard Leakey, in which the renowned paleoanthropologist discussed the results of his fossil-hunting expeditions to Lake Turkana in northern Kenya, Gish says, "We believe that these results support the Biblical record of man's special creation, rather than his origin from an animal ancestry."[2] A few paragraphs later, in initiating a creationist explanation for the primitive nature of cultural materials associated with Neanderthals and other early fossil remains of the modern species *Homo sapiens,* Gish states, "Genesis, Chapter 11, records the fact that there was an early concentration of post-Flood man in the land of Shinar (Babylonia)."[3]

A few minutes' perusal of chapter 6 in Gish's book reveals that the above quotations by no means distort the author's position through removal from context. Indeed, the last four pages of the chapter consist of an exegesis of the biblical account of human origins, as indicated by such headings as "Cain and Abel's Wives—Where Did They Come From?"[4] Insofar as Gish is representative of the creationists (and he is certainly one of their more prolific authors and spokesmen), the group seems committed to the account of human origins contained in the sacred literature of the Judeo-Christian religions. In seeking to explain how we humans got to be the way we are, the creationists do *not* admit "equal opportunity" to the myths of other religions such as the Buddhist, Hindu, or Navajo faiths; only one particular religious tradition is accepted as containing a truthful account of creation. So we are not dealing with a group that simply rejects evolution in favor of *some kind*—whichever seems best—of special creation; we are discussing people who insist upon the literal truth of the biblical account of human origins.

As indicated earlier, the bulk of the creationist literature concerning human origins is not aimed at describing and explaining their own perspective but rather at finding fault with the evolutionary one. Here is a summary of the major creationist criticisms of the evolutionary interpretation of the human fossil record:

1. The human fossil record, like that of all other organisms, is said to lack transitional forms.

2. Fossils that broadly resemble modern humans but differ in important details, such as *"Sinanthropus pekinensis"* (= *Homo erectus*), *"Pithecanthropus"* (= *Homo erectus*), and *Australopithecus,* are merely monkeys or apes that happen to share human traits.[5]

3. No human fossils are older than about six to ten thousand years. When artifacts (such as stone tools) associated with human fossils suggest great antiquity, as in the case of *"Sinanthropus"* and the Neanderthal group of fossils, the rustic appearance of the artifacts is attributed to cultural regression.

4. The data upon which paleoanthropologists base their interpretations of

human fossils present serious problems. These include the difficulty of reconstructing fossil material and the occurrence of specimens that do not fit an evolutionary sequence, such as the 1470 skull discussed later in this chapter.

5. It is implied that paleoanthropologists are unreliable scholars who use inappropriate evidence, are guilty of errors of judgment, and occasionally practice deceit.

The criticisms listed above are intended to serve two broad purposes: (1) to call into question the evolutionary interpretation as a whole, and (2) to refute particular elements of paleoanthropological thinking that conflict with specific parts of the biblical version of creation. These objectives are, of course, mutually supportive, so that some elements of the creationist critique actually serve both.

The first objection, the alleged absence of transitional forms from the fossil record, seems intended as an attack on the evolutionary perspective as a whole. The implication is that the entire notion of evolution *depends upon* transitions; thus, if forms transitional between ourselves and other creatures are lacking, evolution has, ipso facto, not occurred. The alleged absence of human skeletal remains that differ from modern people is also aimed largely at the evolutionary perspective as a whole and is more or less an extension of the argument concerning transitional forms; that is, the alleged absence of primitive human fossils supports the assertion that there never have been forms transitional between ourselves and other creatures. The denial of great antiquity among human fossils is apparently intended to refute evidence that conflicts directly with the biblical account of creation. At stake here are the biblical genealogies that, according to the calculations of a seventeenth-century cleric, would place the date of creation at about six thousand years ago.[6] The last two elements of the creationist critique seem to serve both their broad aims; that is, if the data on human evolution are questionable and the paleoanthropological scholars unreliable, one may reject not only the general evolutionary interpretation of human fossils but also any specific interpretations that conflict with the particulars of the biblical account of creation.

THE EVOLUTIONARY PERSPECTIVE

Perhaps the most important point to make about the evolutionary perspective on human origins is that it does *not* depend upon transitional forms, at least not in the sense meant by creationists. The basic premise of evolutionary interpretations is that humans, like other organisms, have developed from the same ancestral forms that have given rise to their "sister" species in the living world. The key concept in this premise is *transformation* (not transition) *from a common ancestor.* This subtle distinction must be grasped if we are to make informed judgments about the creationist critique. We shall return to this point shortly.

It is widely recognized that the "sister" species of *Homo sapiens,* or modern humans, include the three great apes: orangutan, gorilla, and chimpanzee. The resemblances to humans among these species are readily apparent, even to casual observation of the hands and feet, which contain nails and dermatoglyphs (finger whorls) broadly similar to ours; more striking is the resemblance in their skeletal anatomies, reproductive systems, infants, and biochemistry, especially blood proteins. (Indeed, according to biochemical comparisons, the gorilla and the chimp are nearly identical to human beings, while the orangutan is somewhat more distantly related.) There are, of course, significant differences between humans and great apes, notably in the size of the brain, the teeth, and the parts of the body related to locomotion. In particular, both the absolute and relative size of the brain is much larger in humans than in apes.[7] Also, human teeth are generally smaller, and canines (or eye teeth) differ markedly in size and shape from those of most apes, especially male gorillas and chimps (Figure 8.1). Finally, whereas human limbs are appropriately built for bipedal locomotion, all the apes have limbs specialized for one form or another of locomotion involving all four limbs.

From comparative observations of this type, paleoanthropologists have derived the following expectations about the fossil record. It should contain (1) a common ancestor for apes and humans that exhibits none of the anatomical specializations (for example, in brains, teeth, or limbs) that distinguish one group from the other, and (2) two series of transformations of the common ancestor, one of which approaches the appearance of modern humans while the other becomes progressively more like modern apes.

Several details of these expectations deserve emphasis. One is that the common ancestor for apes and humans is not necessarily expected to be an ape (or, for that matter, a human). Indeed, it is emphatically *not* expected to resemble modern apes, which are highly specialized. Thus, the idea that humans are descended from orangutans, gorillas, or chimps is not relevant to human evolution as understood by contemporary paleoanthropologists.

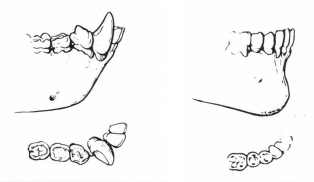

FIG. 8.1. Teeth in apes and humans. Canine-premolar teeth, in side and crown views. In an ape the canine is large, and the first premolar imitates its shape, differing from the second premolar. In humans the canine is small, and both premolars are bicuspid. Used by permission from William Howells, *Mankind in the Making: The Story of Human Evolution,* rev. ed. (Garden City, N.Y.: Doubleday, 1967), p. 107.

Another important feature of evolutionary expectations is that they do not call for a smoothly graded series of forms, or transitions, leading from the common ancestor to the modern creatures. This expectation is incorrectly attributed to the evolutionary perspective by creationists, who then seek to knock down their straw man with the conviction that this will undermine human evolution as a whole. Paleoanthropologists (and others who share their perspective) have long recognized that evolution often proceeds along erratic lines, now emphasizing one part of the anatomy and now another, so that there are no theoretical grounds for expecting a smoothly graded series of transitional forms in the human fossil record.

What do we find in this record? Although it contains substantial gaps, the overall picture is remarkably consistent with evolutionary expectations. The relevant parts of the record come from deposits representing the period from about twenty-five to two million years ago, the Miocene and Pliocene epochs on the geologic time scale.[8] Fossil remains from Africa, Europe, the Near East, and Asia dating to about the earlier two-thirds of this range of time include a large variety of apelike creatures such as the genera *Dryopithecus* and *Ramapithecus* (Figure 8.2), some of which are quite unspecialized and could well contain the common ancestry of modern apes and humans. They lack the large molars and canines of modern apes, the brain development of modern humans, and the locomotor specializations of either modern humans or apes. Some of them are, to be sure, quadrupedal, but they appear to have practiced locomotion more closely resembling the generalized four-legged gait of monkeys rather than the highly specialized knuckle-walking and quadrumanal climbing of modern apes. In short, genera such as *Dryopithecus* exhibit some of the traits common to modern apes and humans, including the form of the

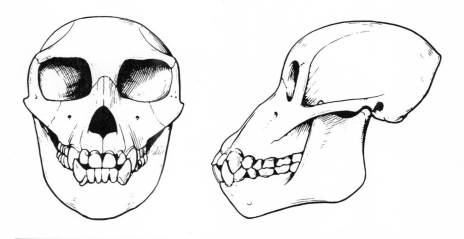

FIG. 8.2.   The skull of *Dryopithecus,* slightly less than one-fourth natural size. Used by permission from William Howells, *Mankind in the Making: The Story of Human Evolution,* rev. ed. (Garden City, N.Y.: Doubleday, 1967), p. 109.

molars, but none of the anatomical specializations that distinguish the two groups of organisms.

However, we cannot confidently determine which, if any, of the known Miocene forms was the common ancestor for apes and humans, largely because fossil material representing each group is exceedingly sparse for crucial periods. Where apes are concerned, the fossil record is impoverished all the way from about eight million years ago to the present. Human remains are very scarce between about eight and four million years ago but are abundantly represented in the fossil record thereafter. The scarcity of ape fossils (perhaps as well as human ones) over the indicated time spans may be related to the ecology of the organisms involved. Modern apes are particularly well suited to life in dense forests, and it may be that their ecological orientation dates as far back as late Miocene times. If so, the humic soils of the forests would have destroyed most of their skeletal remains, thus seriously depleting their fossil record. This does not mean that we will never be able to trace the ancestry of the apes, only that the task of recovering their fossil remains is exceptionally difficult and requires much effort and some good luck.

Although the precise identity of the common ancestor for apes and humans is in doubt, most paleoanthropologists agree that it will probably turn out to be *Dryopithecus* or a closely related form, either among the known species of apelike creatures of the Miocene or among yet-to-be-discovered fossils from that epoch. Despite these uncertainties about the ancestry of apes and humans, the fossil record of the Pliocene epoch contains an array of creatures whose relationships to the two groups are unambiguous and extremely revealing. Most of the bones have been recovered from deposits along the "spine" of Africa — the uplands running the length of the eastern side of the continent, including the great Rift Valley and adjacent regions. Among the more productive sites, from north to south, are the Afar region of northeastern Ethiopia, the Omo Valley of southwestern Ethiopia, the eastern shores of Lake Turkana in Kenya, Olduvai Gorge and Laetoli in north-central Tanzania, and various cave sites in South Africa (Kromdraai, Makapansgat, Swartkrans, Sterkfontein, and Taung). These sites have yielded hundreds of fossils which, though undoubtedly human, contain many indications of an apelike ancestry. Moreover, they include transformations that look increasingly like modern humans as they approach the present. Thus they are the closest thing to transitional forms one might expect to find in the human fossil record, although they are by no means a smoothly graded series.

The earliest species, dating to nearly four million years ago, is called *Australopithecus afarensis*. It differs substantially from modern humans, as is indicated by its assignment to a different genus; *Homo* is, of course, the genus to which we belong. Among the more apelike traits of *A. afarensis* is its small brain size, well within the range of variation for modern apes. But its teeth are intermediate between those of modern apes and humans and, most important, its lower limbs are definitely suited to bipedal locomotion (Figure 8.3), the specialized form of movement that distinguishes humans from apes. Not only do the bones of *A. afarensis* indicate upright posture and bipedal locomotion, but fossilized footprints from Laetoli indicate that the creature practiced this most distinctive type of human behavior.

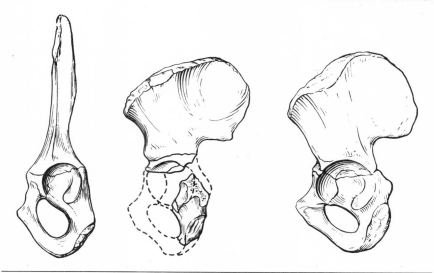

FIG. 8.3. Hipbones, *left,* in side view, of chimpanzee, *Australopithecus,* and man. Used by permission from William Howells, *Mankind in the Making: The Story of Human Evolution,* rev. ed. (Garden City, N.Y.: Doubleday, 1967), p. 122.

The species *A. afarensis* is replaced in the fossil record sometime between about three and two million years ago by two new species: *A. africanus* and *A. robustus* (Figure 8.4). As the names imply, these species appear to have a common ancestry in *A. afarensis* but exhibit substantial transformations. In particular, the canine teeth of *A. africanus* and *A. robustus* are virtually indistinguishable from those of modern humans. *A. robustus* had exceptionally large molars, apparently representing a dietary specialization as yet not well understood, but the molars of *A. africanus* were relatively small, pointing in the direction of modern humans. For this reason, and also because *A. robustus* seems to have become extinct without issue around one million years ago, *A. africanus* is often interpreted as the ancestor for subsequent humans.

By about two million years ago, additional transformations had occurred, particularly in the size of the brain, which had increased by as much as 40 percent, so that it was just beyond the upper end of the range for modern apes and close to the lower end of the range for modern humans. Paleoanthropologists differ as to whether these late Pliocene and early Pleistocene fossils should be assigned to the genus *Homo* or *Australopithecus.*[9] However, the weight of expert judgment seems to favor assigning them to the species *Homo habilis,* which means they are regarded as fully human, that is as belonging to the same genus as modern people. Supporting this judgment is evidence that standardized, albeit crude, stone tools (Figure 8.5) seem to have been routinely manufactured by *H. habilis;* such behavior suggests a major advance in intellectual capacity.

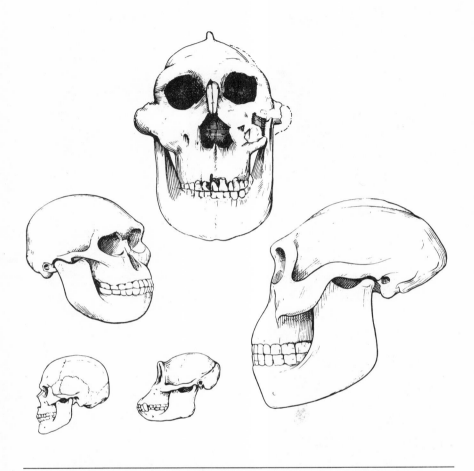

FIG. 8.4. Skulls of *Australopithecus. Top,* "Zinjanthropus" from
Olduvai Gorge, Tanzania, an example of *Australopithecus robustus.*
Left, Australopithecus africanus, from Makapansgat, South Africa.
Right, Australopithecus robustus from Swartkrans, South Africa.
Notable features of *Australopithecus robustus* are the great jaw
height and the short crest along the midline of the skull. All approx-
imately 15 percent natural size. Human and chimpanzee skulls are
sketched in for reference. Used by permission from William
Howells, *Mankind in the Making: The Story of Human Evolution,*
rev. ed. (Garden City, N.Y.: Doubleday, 1967), p. 121.

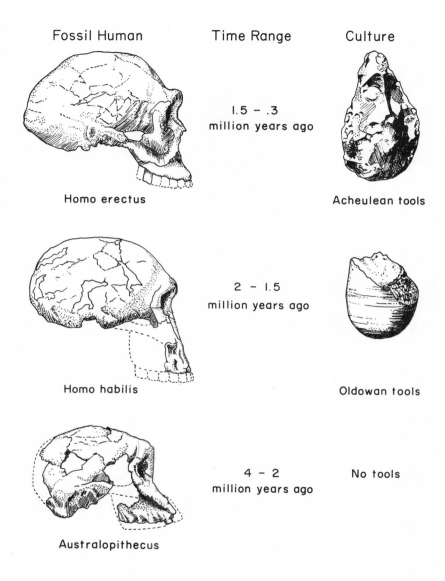

Fossil Human	Time Range	Culture

Homo erectus — 1.5 – .3 million years ago — Acheulean tools

Homo habilis — 2 – 1.5 million years ago — Oldowan tools

Australopithecus — 4 – 2 million years ago — No tools

FIG. 8.5. Progressive enlargement of brain size in human evolution, presumably through selection pressures favoring a large brain, is paralleled by cultural progress, particularly the invention and refinement of techniques of stone tool manufacture.

This is, of course, speculation. But transformations in brain size, whether correlated with intellectual development or not, are the dominant feature of human evolution from the time of the appearance of *H. habilis* onward. Indeed, brain size and related transformations of the skull in fossils from deposits dating to about 1.5 million years ago are so pronounced as to warrant their assignment to a new species, *H. erectus*. This species was cosmopolitan in distribution, inhabiting various parts of the Old World (including Germany, Java, and China), and its skeletal anatomy was essentially modern from the neck down. By about two hundred thousand years ago, it is replaced in the fossil record by the modern species *H. sapiens,* although fully modern humans indistinguishable from living people and belonging to the living subspecies *H. sapiens sapiens* did not appear until at least one hundred thousand years later.

In summary, the human fossil record reveals a series of transformations between about four million and one hundred thousand years ago, involving the development of bipedalism, emergence of modern dental structures, and development (not only enlargement but also complication) of the brain. Moreover, the last of these is more or less paralleled by a series of major advances in human draftsmanship, as indicated by stone tools (Figure 8.5), which may be related to the progressive development of intellect. There is no doubt that the record contains a variety of primitive human forms, and the broad parallels between cultural (that is, technological) and biological development directly contradict the creationists' use of cultural regression to "explain away" the antiquity implied by the crude-looking tools associated with primitive humans. Cultural regression has occurred from time to time, but the major pattern that emerges during the Pleistocene is one of technical progress closely linked with biological development.

Two elements of the creationist critique have yet to be answered: problems in the data relevant to human origins and the unreliability of paleoanthropologists. It is true that the data are fraught with numerous problems, as illustrated in the next section by a particularly notorious example. Many fossils are fragmentary, and the bones are sometimes difficult to restore to their original condition. An added problem is that the dating of human fossils requires highly technical investigations, the results of which are often complex and ambiguous. Finally, there are serious gaps in the fossil record. These are reasons for caution, but they are not reasons for denying that the data have any meaning at all or for simply accepting biblical authority. Most paleoanthropological interpretations of the fossil record are based on the convergence of various lines of evidence, not on isolated pieces of information, and they are generally hedged in terms of the possibility that more and better data may emerge in the future.

As to the alleged unreliability of paleoanthropologists, it is somewhat painful to dignify the charge with a response. Paleoanthropologists are obviously human; this means that they may be vain, they may suffer from lapses of judgment, and they may even occasionally distort the facts to fit their biases. They are also highly trained, disciplined scientists, many of whom perform incredibly arduous, painstaking, and tedious labor for years on end, with meager rewards in money or prestige, simply because of their passionate concern for advancing knowledge. They are, on the whole, comparable with most other groups of scientists in skill, dedication, and integrity.

THE CURRENT STATE OF PALEOANTHROPOLOGY

Every science centers around two separate but closely intertwined lines of endeavor: theory construction and the assessment of theory in the light of data. In addition, a core of solid knowledge forms the base of the discipline. In paleoanthropology, this consists of a large and generally mutually supportive array of evidence indicating that humans have evolved from a common ancestry with the great apes. Some of the relevant paleontological and archeological evidence has already been mentioned, along with evidence from comparative studies of living humans and apes, that is, the neontological evidence. Limited space precludes any systematic effort to demonstrate the validity of the "core of knowledge" in paleoanthropology, and those who wish to pursue the topic further should consult the bibliography at the end of this chapter. However, some attention to the interplay between theory and data is required. Theories are usually generated in response to questions, and two of the fundamental questions in paleoanthropology are, (1) What precipitated the separation of humans and apes? and (2) Has the human gene pool been divided in the past, and if so, how? Theory generated by the first question is obviously crucial to understanding humanity's origins, while theory generated by the second question is important in understanding subsequent evolution.

Early in the twentieth century, the dominant theory regarding the origins of humanity centered on the brain. That is, since comparisons of living apes and humans suggested that the most significant distinction between the two groups was the difference in their intellectual powers, it was theorized that the development of a large and powerful brain in some creatures close to the common ancestry put them on an evolutionary pathway separate from the remaining species. This theory was in trouble from the start, for the fossil remains of *"Pithecanthropus"* (= *Homo erectus*), found in Java late in the nineteenth century, combined a small brain and primitive skull with a virtually modern skeleton from the neck down. If brain enlargement was so important in the origins of humanity, why was it overtaken by developments in parts of the skeleton related to locomotion? The only fossils that really supported the "brain first" theory were the Piltdown finds from England, consisting of a large brain case joined with a primitive jaw, but these were eventually discovered to be a clever forgery, probably perpetrated by an eccentric amateur.[10] However, before the forgery was exposed, there was extensive and often agonized debate in paleoanthropological circles about the meaning of Piltdown and other primitive human fossils in relation to the prevailing theory of human origins.

Even after the exposure of the Piltdown fraud, the theory of brains first was not entirely discarded. Instead, it was incorporated into a more complex "feedback" theory, in which the use of tools, bipedalism, and intellectual development were viewed collectively as the initial step in human evolution and each was regarded as reinforcing the other. Briefly, the argument was that toolmaking required freeing the hands from locomotion and put a premium on intellectual development, which enhanced the ability to make tools, which further stimulated the liberation of the hands through bipedalism, and so on. The theory was supported by evidence from Olduvai Gorge indicating that at least some members of the earliest human genus, *Australopithecus,* may have used stone tools. But the theory has been punctured by the recent discovery of *A. afarensis,* which is a fully bipedal human that predates the earliest evidence for tool manufacture by at least one million years. Thus current theories of human origins assume that bipedalism was the triggering factor and seek to explain the importance of bipedal locomotion.

Returning to the question of how, or whether, the human gene pool has been separated over time, we find that the response has been divided between two major competing models (or theories) for just about as long as paleoanthropology has existed. One model, based largely on observation of the living "races," postulates separate gene pools, and thus separately evolving human lineages, for some if not all the time during which human evolution has occurred.[11] This can be called the Multilinear Model. In its extreme form, this theory attributes a separate evolutionary history to each "race" all the way back to the common ancestor for all of them and is known as the Candelabra Model (Figure 8.6). The competing theory stipulates that all varieties of humans have always shared in a common gene pool, so that all fossil humans are direct ancestors (not merely collateral ones) of all living humans. In its simplest form, this is a Unilinear Model of human evolution. However, a somewhat more elaborate variant, known as the Hatrack Model (Figure 8.6), admits the possibility of minor branching in the gene pool, with extinction as the ultimate fate of the branches.

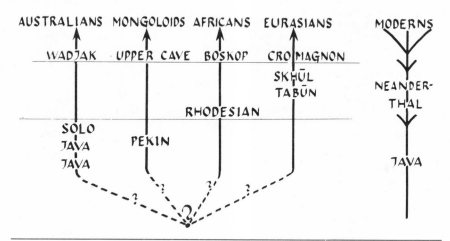

FIG. 8.6. Graphic representations of different theories of human descent. *Left,* the Polyphyletic or Candelabra Model. *Right,* the Unilinear or Hatrack Model. Used by permission from William Howells, *Mankind in the Making: The Story of Human Evolution,* rev. ed. (Garden City, N.Y.: Doubleday, 1967), p. 241.

Each of these theories has its contemporary advocates. Among those drawn to the Multilinear Model is Richard Leakey, director of the field investigations at Lake Turkana in Kenya. In 1972 Leakey discovered a fossil designated ER 1470, the notorious 1470 skull, which was dated close to three million years before the present, well within the time range of *Australopithecus,* but which had a much less primitive appearance and a much larger brain. This discovery seemed to support the Multilinear Model, for it implied that a creature closely related to ourselves, and presumably ancestral to us, had coexisted with a more primitive, collateral form for something in excess of one million years. However, the original dating of the 1470 skull was problematical, since it was based on the average value for a very wide scatter of potassium-argon dates.[12] Subsequent geological and geophysical studies have yielded a revised and apparently secure date of about two million years ago. This puts 1470 in the range of time for *Homo habilis,* which is almost certainly what it is. But, as we have seen, *H. habilis* appears to be the descendant of *Australopithecus,* so the 1470 skull can no longer be used as evidence in favor of the Multilinear Model.

And so it goes. Theories are formulated and then challenged in the light of new data. Meanwhile, the core of knowledge continues to grow. Nothing is certain, but our understanding of human origins continues to deepen through use of the scientific method in paleoanthropology.

FOR FURTHER READING

BRACE, C. L., AND MONTAGU, ASHLEY. *Human Evolution: An Introduction to Biological Anthropology,* 2nd ed. New York: Macmillan, 1977.
   An exceptionally well-written textbook, which includes detailed accounts of early fossil finds such as *"Pithecanthropus"* and Piltdown.

CLARK, W. E. LE GROS. *The Antecedents of Man.* New York: Harper & Row, 1959.
   A highly technical but masterful comparative anatomical study of humans and their nearest relatives among the mammals, including the living apes.

CLARK, W. E. LE GROS. *Man-Apes or Ape-Men: The Story of Discoveries in Africa.* New York: Holt, Rinehart & Winston, 1967.
   An excellent description of fossils representing *Australopithecus* and the evidence indicating that the genus is human.

COON, CARLETON S. *The Origin of Races.* New York: Alfred A. Knopf, 1962.
   A somewhat dated but exhaustive description of the fossil record for human evolution. Contains an interesting though ultimately unconvincing argument for the Candelabra Model.

JOHANSON, DONALD, AND EDEY, MAITLAND A. *Lucy: The Beginnings of Humankind.* New York: Simon & Schuster, 1981.
   A thoroughly engaging personal account of the discovery of *Australopithecus afarensis,* with comments on the theory of initial human divergence from common ancestry with the apes. Written for a lay audience.

PFEIFFER, JOHN E. *The Emergence of Man,* 2nd ed. New York: Harper & Row, 1972.
   Similar in content to the book by Brace and Montagu but emphasizes the role of behavior in human evolution.

SHAPIRO, HARRY L. *Peking Man.* New York: Simon and Schuster, 1973.
   A lively and informative account of the discovery, interpretation, and loss of the *"Sinanthropus"* fossils. Excellent source of information on one of paleoanthropology's most controversial finds.

[9]

Thermodynamics:
The Red Herring

HUGO F. FRANZEN

THIS chapter deals with the application of the laws of thermodynamics to concepts of stellar and biological evolution. These are difficult areas of controversy because a great deal is not known about the origins of our universe and life on our planet. However, thermodynamics is well understood, and all sides agree that the laws of thermodynamics are valid. These laws state that in any process energy is conserved but degraded. Degradation of energy refers to a loss in the ability to do work. The scientific creationists have introduced these laws into the evolution-creation debate to present arguments relating to the origin of the universe and the developing complexity of life forms through biological evolution.

A fundamental difference between the goals of the creationists and those of scientists should be recognized at the outset. Among the goals of the creationists is the explanation of origin and termination, time and space, matter and energy, life and death, meaning and purpose. In contrast, a major goal (perhaps the principal one) of scientists is to discern relationships among natural phenomena. Examples are relationships among force, mass, and acceleration, among force, charge, and field strength, or among heat, work, and energy. Science, then, is concerned with pursuit of verifiable knowledge about the world in which we find ourselves.

The scientist does not consider it a failing that science does not at all explain the origin of the universe or completely explain the origin of life. A lack of explanations for some phenomena is viewed by scientists as a challenge to seek further relationships and insights rather than as a defeat of the scientific method. Scientists recognize that there are many important phenomena for which no acceptable interpretations in terms of fundamental interrelationships are currently available.

The scientific creationists attempt the unification of the disparate goals of scientists and creationists. The result is that they have overextended and misinterpreted the laws of science to "prove" points of creationist doctrine and, in the last analysis, have deserted science in an appeal to supernatural explanations of natural phenomena. The scientific creationists use the laws of thermodynamics, for example, to argue that the universe was created at a specific time by a divine creator. The principal scientific objection to this argument is simply that it is beyond our ability to reach conclusions about the origin of the universe given our current level of understanding.

Science, of course, will continue to provide insights into the possible meaning of an "origin" of the universe and into the relationship of this meaning to our current concepts of time and space. Along the way, some will use science as they perceive it to achieve nonscientific ends. Among them are the scientific creationists. It is appropriate that the scientific community ask them not to misrepresent science, and this chapter is one scientist's contribution to that request.

BASIC THERMODYNAMICS AND TIME

Thermodynamics is a basic science that is widely, if not universally, accepted because no claims of violation of its laws have withstood the test of time. Over the years scientists have attempted to devise schemes to violate the laws of thermodynamics and, one by one, they have fallen by the wayside. This effort has been thorough and convincing. Thermodynamics today can be said to be one of the best established of our sciences.

It is generally accepted that the laws of thermodynamics provide a natural direction to the flow of time, at least in the space and time that have been observed by man. According to thermodynamics, time flows in the direction in which there is a decrease in the total capacity of the energy in our universe to do work. The total energy content of the known universe is constant, but the ability of that energy to do work is degraded with time.

For example, the sun is currently the principal energy source for the earth, but it will not provide radiant energy forever. The solar processes will alter with time and eventually cease. The energy produced by the sun will have been radiated into space, and that energy will no longer exist in a form suitable for doing work. The sun's energy content is degraded with time, and so, too, is the total energy within the universe. At least, so it appears to us at this time.

Scientific creationists claim that this degradation points to an eventual running down of the universe. That is, when the total energy content of the universe has been degraded, all processes will cease and the universe will come to an end. Such an end points back to a beginning, and this, according to the creationists, was divine creation.

The first weakness in this argument is that the supposed act of creation defies the very laws of thermodynamics upon which the argument is based, for the proposed process involves the creation of energy in an undegraded form. While there is a problem at some point of reconciling the apparent degradation of useful energy (that is, increase in entropy) with the constant energy content of the universe if the process is followed backward in time, it requires a leap of faith to conclude that this problem is solved only by accepting a supernatural event.

The alternative to this leap of faith is to recognize that our present ability to reason meaningfully about the origin or end of the universe is not adequate to the job. Just as the spatial limits of the earth were incomprehensible before the earth was recognized to be a sphere, and the spatial limits of the universe were incomprehensible before space-time was recognized to be a hypergeometric surface, so the beginning and end of time are currently beyond our comprehension. The postulate of a divine creator simply begs the issue and skirts the difficult physical questions concerning the creation of space, time, matter, and energy that must be answered before the thermodynamic problems can be understood.

It should be recognized that the concept of a beginning before which there was no time, no space, no energy, and no matter, leaves at least as many unanswered questions as acceptance of the current incomplete understanding of the origin of the universe. These unsettling scientific questions about the origin of matter and energy and of space and time are not settled by postulating divine creation; they are simply ignored by recourse to supernatural supposition.

Although it is not possible scientifically to refute the idea that the universe was created by a supernatural power at the beginning of time, it is possible to say that this idea does not follow from thermodynamic arguments and, furthermore, that such an explanation leaves many important scientific questions about the nature of the event unexplored.

THERMODYNAMICS AND MOLECULAR SYSTEMS

The scientific creationists also apply thermodynamic arguments to the consideration of molecular systems such as those in living organisms. Scientists know a great deal more about this and can give more specific responses. First it is necessary to provide some relevant background details of thermodynamics.

Thermodynamics is a science that deals with the possibility or impossibility of natural processes. The foundation of thermodynamics is contained in two laws. The first deals with the changes in energy that accompany processes. It states that although the form of energy (that is, heat or work) changes in a process, the total amount of energy remains constant. An example of an impossible process, then, is one that creates or destroys energy.

The second law deals with changes in a priori probabilities that accompany processes and leads to the definition of the thermodynamic property called entropy.[1] According to the second law, when all the related changes in a system and its surroundings are taken into account, processes that either do occur or can be made to occur always involve changes from less probable to more probable states. Entropy is a measure of this total probability. Thus the second law states that possible processes are those in which the total entropy of the universe increases. It can be proved that this also means that energy is degraded in possible processes. For this proof, see the references listed at the end of this chapter.

According to the laws of thermodynamics, then, possible processes are those for which the energy content of the universe remains constant and the entropy content of the universe increases. These laws can readily be applied to chemical processes by comparing the entropy change for the process with the entropy change for the rest of the universe. This latter quantity — the entropy change of everything except the reaction itself — is given by the energy evolved in the form of heat during the reaction divided by the temperature at which the reaction occurs. The sum of these two entropy changes is the total entropy change of the universe. According to the second law, this sum is positive for a possible process.

Before considering a concrete example, we should become familiar with the techniques of dealing with numbers of enormous magnitude in order to appreciate the quantities involved. These quantities are computed from a knowledge of the number of states available to a system. In the case of a system of molecules that cannot exchange matter with the surroundings, the number of ways the system can exist with a fixed temperature and volume is the number of states accessible to the system. For an atomic gas at room temperature and atmospheric pressure, this number is obtained by counting the different distributions of velocities of the atoms consistent with the given temperature and volume. For a system of any reasonable size, say one cubic inch, this number is *incredibly* large.

In order to deal with numbers of the size required for a discussion of system probabilities, scientific notation is usually used, and 100 is written 10^2, 1,000 is 10^3, and 1,000,000 is 10^6, and so on. Thus the exponent (n) of 10^n is the number of zeros following 1. The number of molecules in one cubic inch of water at room temperature is, then, approximately 10^{23}, and the number of states accessible to this system is approximately $10^{10^{24}}$ (that is, 1 followed by 10^{24} zeros — if it takes you ½ second to write a zero, it would take you 2×10^{14} lifetimes of 100 years just to write this number in conventional notation).

Consider a very simple chemical reaction such as the association of oxygen atoms to form one mole of molecular oxygen (O_2): $2O = O_2$.[2] At present, the availability of extremely reliable data makes it a simple matter to calculate the number of states accessible to the oxygen atoms for such a reaction (the number is $10^{10^{25}}$), as well as the number of states accessible to the molecular oxygen, also at room temperature and atmospheric pressure (the number is $10^{6 \times 10^{24}}$). The ratio of the two numbers is $10^{4 \times 10^{24}}$ (or 1 with 4×10^{24} zeros after it). Thus there are enormously more states accessible to the atomic oxygen than the molecular oxygen and, other things being equal, it is much more probable that oxygen would be found in a dissociated (atomic) state than in an associated (molecular) state. Considering only the a priori probabilities, as above, it would follow that oxygen molecules would be present to only eight parts in ten million, according to straightforward calculation from these numbers. However, the oxygen in air, upon which we depend for our very lives, is overwhelmingly molecular O_2 and contains virtually no dissociated atoms.

The discrepancy between what is observed (vast preponderance of molecular O_2) and what is calculated (vast preponderance of atomic O) has its origin in the phrase "other things being equal." Other things are not equal. Recall that the criterion for the tendency of a process to occur is the *total* entropy change, that is, total entropy change of the system and its surroundings (the rest of the universe). In the erroneous calculation, the entropy change of the surroundings was neglected. As stated above, the correct result can be obtained only by comparing the heat energy released by the system divided by the temperature with the entropy change of the system. The incorrect result that atomic oxygen is favored over molecular oxygen was obtained by considering only the system probabilities and thus only the system entropy change.

In considering the entropy change of the rest of the universe, we are led naturally to inquire about the energy change accompanying a chemical reaction. We are aware of such energy changes in our daily lives in combustion processes, in batteries, and particularly in the energy source for life itself, metabolism. Such chemical energy changes can be analyzed in terms of the energy released upon formation of a chemical bond, or "bond energies." That is, one molecule of oxygen is at a lower energy than two atoms of oxygen; therefore a decrease in the energy of the system accompanies the reaction to form molecular oxygen. This energy change must be considered in order to reach the correct conclusion regarding the tendency for the reaction to occur. When it is considered, thermodynamics predicts that molecular oxygen will predominate by an overwhelming amount (10^{20}:1), and this predominance of molecular oxygen is observed in nature. As illustrated in Figure 9.1, when oxygen atoms form chemical bonds characteristic of molecular oxygen, two entropy changes occur. First, because the number of O_2 molecules is less than the number of O atoms, the entropy within the container decreases. Second, the heat given off during the formation of the chemical bond is transmitted to the universe surrounding the container, thus increasing the entropy of the surrounding universe. In Figure 9.1 the total entropy change (372 entropy units) is positive, and therefore the reaction is thermodynamically possible.

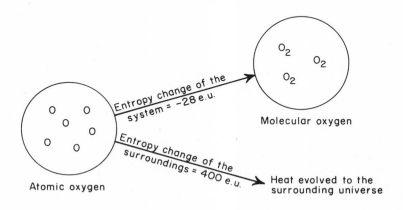

FIG. 9.1.   Entropy changes involved in the chemical bonding of oxygen atoms into oxygen molecules. An e.u. is an "entropy unit," a measure of entropy. Values cited are for room temperature.

All natural processes and, in particular, all chemical reactions involve competition between the tendency to achieve the a priori most probable configuration of the system (that is, the configuration that maximizes the number of states accessible to the system and therefore the entropy of the system) and the state with the lowest energy (which maximizes the entropy of the rest of the universe). The release of heat energy with the formation of a chemical bond thus can be seen to have a profound influence on the world in which we live. Without the energy of the chemical bond, important molecular systems would not exist. Other things being equal, the probability of the formation of molecules is so fantastically small that without chemical bond energies their occurrence would be negligible.

On the other hand, the energies of chemical bonds drive molecular systems in a direction counter to that suggested by the extraordinarily large a priori system probabilities for dissociation and produce a world of incredible molecular diversity. As far as we know, molecules—from the smallest (H_2, N_2, O_2) to the largest (proteins and DNA)—always form in accordance with the laws of thermodynamics.

Furthermore, the increase in the underlying molecular complexity accompanying the early stages of evolution was reflected in an increase in the complexity of the life forms themselves. That is, from the chemical point of view, the evolution of complex living organisms occurred as the result of changes in the molecular systems upon which the organisms depended for their reproduction, growth, metabolism, and other life processes. Since it is at the molecular level that the second law of thermodynamics straightforwardly applies, it follows that evolutionary concepts are consistent with the laws of thermodynamics.

In contrast, scientific creationists argue that the complexity on earth (for example, the molecules of living organisms) is so improbable that it is thermodynamically impossible and thus requires divine intervention as a necessary agent in its formation. Some of the concepts developed above cast serious doubt on this assertion.

First, as we have seen, chemical processes do not occur in such a way as to be in agreement with the consideration of system probabilities alone. Even for the simplest molecular systems, the probabilities against formation are enormous. By what means, the scientist asks, does the scientific creationist differentiate between improbable events that require divine intervention and those that do not? In short, is divine intervention required to form molecular oxygen? If not, why not? The formation of O_2 is overwhelmingly improbable according to the line of argument adopted by the creationists.

Second, what about the need to consider bond energies in the formation of complex molecules? As already pointed out, if the bond energies involved are neglected, the result is an incorrect conclusion. Thus the central problem of the creationists' arguments concerning the formation of complex molecules is that they consider the entropy change of the system alone and neglect the entropy changes in the remainder of the universe. Yet in the case of molecules with substantial bond energies and moderate temperatures (see Figure 9.1 for values for oxygen), the entropy change of the surroundings is by far the larger of the two entropy changes. The creationists' assertion that an intelligence is required to overcome the probabilities is entirely without thermodynamic basis.

The point under discussion is so important to the understanding of the application of thermodynamics, and is so poorly understood, that it bears some repetition. When thermodynamics is applied to the consideration of the possibility of a process, it is essential that the total entropy change of everything that changes in any way — that is, the total entropy change of the system plus its surroundings or that of the total universe — be considered. According to the second law of thermodynamics, this total entropy must increase during a possible process. Nonetheless, the entropies of parts of the universe may decrease as long as the entropies of other parts increase by a greater amount. From the statistical (probabilistic) point of view, this means that the total universe is moving toward an ever more probable configuration (as far as we know) but that in the process parts of it can be evolving into less probable states. A part of the universe can, in agreement with the laws of thermodynamics, increase in organization and complexity as long as the total is decreasing in these qualities.

It is therefore consistent with thermodynamics for a spontaneous process to occur in which a system proceeds to a less probable state (decreases in entropy). Such a process cannot occur unless the increase in the entropy of the remainder of the universe is such that the total entropy increases. However, the scientific creationists assert that the increase in complexity that accompanies evolutionary processes runs counter to the second law. This is like looking in a refrigerator and arguing that because the inside is colder than the surroundings, the second law has been violated. Such an assertion neglects the increase in entropy accompanying the processes producing the electrical power driving the refrigerator motor. In the case of many processes on earth, the solar processes providing the power to drive the terrestrial processes cannot be neglected. There is more than enough entropy produced in the sun to explain the spontaneity of processes occurring with a decrease in entropy here on earth.

NET PROCESSES AND MECHANISMS

Physical chemists have for many years recognized the importance of the distinction between a net process (described by the initial and final states) and the mechanism of a process (described by the path along which the process occurs). The former is accepted as the province of thermodynamics and the latter as a branch of science called kinetics.

A net process, from a thermodynamic point of view, has a certain tendency to occur. This tendency can be quantified by calculating the sum of the entropy change of the system and the heat evolved by the system divided by the temperature, as already described. Even if a process is found to have a tendency to occur, there remains the question of the rate at which the process occurs, or whether it occurs at all. It may be, and for many processes it is, the case that a process is spontaneous (that is, possible) from the thermodynamic point of view but does not occur at all because no path exists that allows it to proceed. In these instances a catalyst is a substance which, by virtue of its presence, can provide a path that will allow the process to occur and yield the final product. The final product is determined by thermodynamics, and the catalyst provides *only* the mechanism that provides a pathway for the process.

As a simple example, consider a mixture of hydrogen gas and oxygen gas at room temperature. According to thermodynamics, the gases should react to form water: $H_2 + (\frac{1}{2})O_2 = H_2O$. They will do so explosively if a spark is struck in the mixture. Recall that hydrogen mixed with oxygen is a rocket fuel. If the mixture is contained in a glass vessel in the absence of a spark, it will remain indefinitely without reacting. Platinum metal is a catalyst for this reaction. If platinum is introduced into the vessel, the H_2 and O_2 will react to form H_2O on the surface of the platinum. The reaction yields exactly the final product predicted by thermodynamics. Physical chemists stress the important principle that a catalyst affects only the rate (kinetics), not the net process (thermodynamics).

The complex molecules of life (RNA, DNA, enzymes) are catalysts and are themselves formed in catalyzed reactions. The catalysis cannot affect the net process without violating the laws of thermodynamics. If complex molecules of life are forming at the present time (and they certainly are in every living organism), they must form via a possible net process according to the laws of thermodynamics. If, as the scientific creationists contend, a divine creator is required on thermodynamic grounds for the original synthesis of the complex molecules, then such a creator is required for each and every synthesis. It is not consistent with thermodynamics to suggest, as the creationists have, that the creator provided catalysts that made otherwise impossible processes possible. Such a suggestion confuses kinetics and thermodynamics and, in the process, denies the validity of thermodynamics.

The implication is that if thermodynamics requires the intervention of a supernatural agent to originate life, it requires the continued intervention of that agent to sustain life. There would then be a sustained and repeated violation of the laws of thermodynamics in life processes. It is, of course, possible to accept this notion. It is not possible to accept it and thermodynamics as well.

Although the above arguments do not disprove the creationists' thesis that the universe and life were created by a supernatural force at some point, they do indicate a number of logical and scientific problems in the argument that creation by a supernatural being is somehow required by the laws of thermodynamics. The arguments of scientific creationists are erroneous in their applications of thermodynamics to evolutionary concepts. Moreover, the creationists have tried to use thermodynamics, which deals only with net processes, to analyze evolution, which is in its essence a mechanism for the development of life as we know it. Since what is principally unknown and controversial is the mechanism for the initiation of life and not the net process, it follows that thermodynamics is truly the red herring of the evolution-creation debate.

FOR FURTHER READING

AUGUST, S. W., AND HEPLER, L. G. *Order and Chaos.* New York: Basic Books, 1967.
A nonmathematical treatment of the concepts of thermodynamics and their application to processes of general interest.

BENT, H. A. *The Second Law.* New York: Oxford University Press, 1965.
An introduction to the basic ideas of thermodynamics with a number of lucid examples of applications to familiar processes.

FERMI, E. *Thermodynamics.* New York: Prentice-Hall, 1937.
A discussion of thermodynamics by one of the great physicists of our century. Rigorous and insightful.

PLANCK, M. *Treatise on Thermodynamics.* New York: Dover, 1945.
A discussion of thermodynamics by one of the great physicists of the past. Rigorous and insightful.

ZEMANSKY, M. W. *Heat and Thermodynamics,* 5th ed. New York: McGraw-Hill, 1968.
A standard in the field, the traditional first text in thermodynamics. Clear and rigorous.

III

The Realm of Religion

[10]

Creation Belief in the Bible and Religions

PAUL HOLLENBACH

W E turn now to the question of how modern scholarship views traditional belief in creation among the religions of the world. The
character of the creation stories and of creation belief is best illustrated in the various religious texts, some of which are examined in this
chapter. Our discussion focuses on the opening three chapters of Genesis, as
they are the central scriptural creation text for both Judaism and Christianity.
Examination of these texts also makes possible some observations about creation belief in the context of the modern controversy about creation and
evolution.

Approaching the controversy through the literature of religions may seem
superfluous in view of the well-known assertion of some creationists that their
creationism is "scientific" and has nothing to do with religion, particularly not
with the Genesis narratives. This assertion is belied by much creationist belief,
not only on the popular level but also on the more sophisticated level. It goes
without saying that in the mind of the general public the creator hypothesized
by scientific creationists is the creator God of the Bible. And at least some professional creationists clearly make the same identification. For example, each
issue of the *Creation Research Quarterly* contains a statement of beliefs to
which all members of the Creation Research Society must subscribe. The first
two are:

> 1. The Bible is the written Word of God, and because we believe it
> to be inspired thruout, all of its assertions are historically and scientifically
> true in all the original autographs. To the student of nature, this means
> that the account of origins in Genesis is a factual presentation of simple
> historical truths.
> 2. All basic types of living things, including man, were made by
> direct creative acts of God during Creation Week as described in Genesis.
> Whatever biological changes have occurred since Creation have accom-
> plished only changes within the original created kinds.

Here the God of Genesis and of the whole Bible is clearly equated with the creator of all living things, whose existence creationists claim is proved through their particular form of scientific and philosophical reasoning. That some creationists really do make this identification was recently indicated by an article in the *Creation Research Quarterly* in which the author, after discussing the problem of whether creationists base their views on the Bible, concludes that "the dependence our theories and hypotheses have upon Biblical pronouncements—and our interpretation of them—[is] extremely clear."[1] Therefore, an examination of this biblical basis for creationism not only seems justified and relevant to the current controversy, but it also will help us to understand the character of contemporary creationism as a religious and social phenomenon.

It is important to note in the above quotation the author's careful addition of the phrase "and our interpretation of them" (that is, the biblical texts). This phrase is added quite properly, for there is no possible reading of a text that is not an interpretation. This includes, of course, the Creation Research Society's own interpretation of Genesis indicated in the two statements of belief just presented. Our purposes in this chapter will be served if we begin our analysis with this representative and official interpretative statement, our goal being to test the cogency of this kind of interpretation, which asserts that "Genesis is a factual presentation of simple historical truths," that "its assertions are historically and scientifically true."

Initially, we want to see whether a reading of the first chapter of Genesis as a factual, scientific description of events makes good sense. But this must be a careful, literal reading based on the best linguistic and historical scholarship available. Only in this way can we avoid reading into the text meanings that we may wish to see but really are not there. Hence, unlike most creationists, we must use the methods and results of modern biblical linguistic and historical scholarship as they are found, for example, in the writings of members of the Society of Biblical Literature, the main professional society for North American biblical scholars.

We offer a careful translation of the early verses of Genesis, along with observations pertinent to the issues of this book.[2] First consider verses one through three:

> When God set about to create heaven and earth (or: At the beginning of God's creating heavens and earth)—the earth (world) being then formless and empty, with darkness over the surface of the deep seas, with only the breath of God (or, an awesome wind) sweeping over the water—God said, "Let there be light." And there was light.

The technical reasons for constructing these three verses in this form are thoroughly explained in E. A. Speiser's book on Genesis.[3] It should be noted that verse one is a subordinate clause that depends on the main clause appearing in verse three, verse two being a parenthetical observation.

What then do these verses say? Some aspects of the early events are clear, whereas others are not. It is clear that when God began his creative activity some entities already existed—darkness, water, and air—which were, in a sense, very raw materials since they were shapeless, without thingness. Yet they must have had some minimal shape, since the waters (deep seas) evidently had a surface over which wind could pass and above which darkness could be. Further, it is clear that God first created light. Since it was created first, perhaps it is of special importance. Let us keep that in mind for later reflection.

But now, what is not clear? We don't know, for example, what the phrase "heavens and earth" refers to, since in some sense the earth or world already existed, as verse two says—the same Hebrew word for earth appears in verses one and two. How could God create (heavens and) earth when earth already existed? Where did this preexisting darkness, water, and air come from? We are not told at all. What was the shape of the surface of the waters? Round? Flat? Square? If flat, were there edges? If so, what held the water in? If round . . . well, it can be seen that numerous questions arise because so much about the beginning of God's creating activity is left unspecified. One particular question leads us to verses four and five: When the light appeared, what happened to the darkness? Was it totally obliterated or was it somehow pushed aside? Verses four and five apparently try to answer this question.

> God was pleased with the light that he saw, and he caused a division between the light and the darkness. God called the light day, and he called the darkness night. And there was evening and morning—day one.

In these verses we learn that the darkness was not obliterated. Rather, God evidently separated it from the light he had just created; this was not a spatial but a temporal separation, since he called the light day and the darkness night. Does this mean, then, that there was already day and night in the twenty-four-hour sense? Although the story does not say, scholars suggest that the closing phrase "evening and morning" refers to a twenty-four-hour day, since this phrase is a common Hebrew idiom for an ordinary twenty-four-hour day. Further, the text suggests that twenty-four-hour days initially existed without sun, moon, and stars, which were created only on the fourth day. How are we to understand these early twenty-four-hour days (days one, two, and three) without sun, moon, and stars? Puzzles seem to be piling upon puzzles as we try to understand how Genesis can be interpreted as a scientific description of the earliest events in the formation of the world.

In verses six through eight, God turns from the original darkness of verses four and five to the primal waters:

> God said: "Let there be a firm sheet in the middle of the waters so as to cause a division between the waters." And it was so. God made the firm sheet, and it divided the waters below it from the waters above it. God called the firm sheet sky. And there was evening and morning—day two.

Here God constructs a waterproof sheet of material inside the deep seas. The statements in these verses together with the meaning of the Hebrew word used in II Samuel 22:43, Exodus 39:3, and in other parts of the Bible, show that we must understand that what God eventually calls sky is a firm sheet. It has to hold water. Exactly how it does this, however, is no clearer than the explanation in verses four and five of how God caused a division between light and darkness. Nor are we told what holds up this firm sheet or on what it rests. The only reason that we do not imagine the firm sheet to be in a vertical, horizontal, or some other position in the deep seas is that God called the sheet sky. Because of the meaning we attach to this word, we imagine it to be above the waters, the air (the fearsome wind of verse two) being present between it and the waters; and presumably day and night are somehow still in the air.

Nothing further is said about the waters above the sky,[4] but verses nine through thirteen discuss the region below the sky:

> God said, "Let the waters under the sky be gathered into a single area, that dry land may be seen (or, may appear). And it was so. God called the dry land earth, and he called the gathered waters seas. God was pleased with what he saw, and he said, "Let the earth burst forth with growth: plants that bear seed, and every kind of fruit tree on earth that bears fruit with its seed in it." And it was so. The earth produced growth: various kinds of seed-bearing plants, and trees of every kind bearing fruit with seed in it. And God was pleased with what he saw. And there was evening and morning — day three.

Now the waters beneath the air are concentrated so that there may be dry land. We are told neither where this dry land came from, nor whether the dry land was raised so the waters would not cover it again. Or was not the dry land under the waters? Again, we cannot be sure about any of these matters.

Now God turns from organizing the earth to organizing the heavens in verses fourteen through nineteen:

> God said: "Let there be lights in the firm sheet of the sky, to distinguish between day and night; let them mark the fixed times, the days and the years, and serve as lights in the firm sheet of the sky to shine upon the earth. And it was so. God made the greater lights, the greater one to dominate the day and the lesser one to dominate the night — and the stars. God set them in the firm sheet of sky to shine upon the earth, to dominate the day and night, and to cause a division between light and darkness. And God was pleased with what he saw. And there was evening and morning — day four.

We notice clearly that the sun, moon, and stars are fixed in the firm sheet. But how we are to understand their motion is not indicated at all. Further, they cause the division between day and night; yet that has already been caused in verse four, that is, on the first day. Why this repetition? And how was the earlier division made without the greater lights? Do these lights give light in addition to the light of the first day? Or do they now somehow become the sources of all the light there is? Again, the story leaves us puzzled if we try to read it as a literal description of beginning events in our solar system.

There is no need to continue the translation of Genesis in order to make the main point, namely, that if Genesis is interpreted historically and scientifically as creationists wish to do, then we must say that the narrative totally fails to describe the universe. For not only is it a baffling, puzzling, incomplete description of the creation of a world, but insofar as we can determine, the narrative is flatly contradicted by modern cosmology. Thus our effort to interpret the biblical creation story in a creationist manner ends in a negative conclusion. We are forced to conclude that the story, insofar as we can understand it, is false from a scientific point of view.

This conclusion also makes the very belief of creationists — that Genesis gives us a true scientific account of origins — puzzling indeed. How can one explain their special commitment to the biblical narratives? Probably the only explanation is that their commitment is a religious one, as shown by the fourth statement to which members of the Creation Research Society must subscribe. It is explicitly religious and Christian since it requires members to affirm faith in Jesus as "Lord and Savior." One must also conclude that creationists do not participate in modern critical biblical scholarship, or else they would see quickly, as we have seen here, that the cosmology in the Genesis narratives is completely at variance from the one accepted today, and that it coheres generally with ancient Near Eastern cosmologies.[5] Thus it appears that creationists approach the Bible largely in a religious, not in a critical, manner. It is their *religious* belief that Genesis is true scientifically and historically.

All these considerations lead one to ask whether there is another way to read and understand the Genesis narratives that makes good sense. Let us now consider one such interpretation, which is supported by modern biblical scholars and those studying religion in the humanities and social sciences. Perhaps the crucial activity of this modern scholarship has been the collection of numerous creation stories from practically all the world's religions and cultures. The extended work on this material by scholars of many disciplines over the last century or so has made the character and purpose of these creation stories clearer and clearer, and they are still being studied in order to understand them better. Consider, for example, one or two representative creation stories out of the dozens that are available. The Omaha Indian story comes from a culture totally separate from our own, while the Babylonian story comes from the same cultural setting as the biblical one.

Omaha cosmogony (as told by an Omaha Indian):

> "At the beginning," said the Omaha, "all things were in the mind of Wakonda. All creatures, including man, were spirits. They moved about in space between the earth and the stars (the heavens). They were seeking a place where they could come into bodily existence. They ascended to the sun, but the sun was not fitted for their abode. They moved on to the moon and found that it also was not good for their home. Then they descended to the earth. They saw it was covered with water. They floated through the air to the north, the east, the south, and the west, and found no dry land. They were sorely grieved. Suddenly from the midst of the water uprose a great rock. It burst into flames and the waters floated into the air in clouds. Dry land appeared; the grasses and the trees grew. The hosts of the spirits descended and became flesh and blood. They fed on the seeds of the grasses and the fruits of the trees, and the land vibrated

with their expressions of joy and gratitude to Wakonda, the maker of all things."

Mesopotamian cosmogony (as summarized by Mircea Eliade):

The long Babylonian creation epic "Enuma Elish" ("When on high"), so called from the first two words of the poem, narrates a chain of events beginning with the very first separation of order out of chaos and culminating in the creation of the specific cosmos known to the ancient Babylonians. As the gods are born within the commingled waters of their primeval parents, Apsu and Tiamat, their restlessness disturbs Apsu. Over Tiamat's protests, he plans to kill them; but the clever Ea learns of his plan and kills Apsu instead. Now Tiamat is furious; she produces an army of monsters to avenge her husband and wrest lordship from the younger generation. The terrified gods turn to Ea's son Marduk for help. Marduk agrees to face Tiamat, but demands supremacy over them as compensation. They promptly assemble, declare him king, and send him forth, armed with his winds and storms. The battle is short; the winds inflate Tiamat's body like a balloon and Marduk sends an arrow through her gaping mouth into her heart. He then splits her body, forming heaven and earth with the two halves. After putting the heavens in order, he turns to Ea for help in creating, out of the blood of Tiamat's demon-commander Kingu, the black-haired men of Mesopotamia. The poem concludes as the gods build a temple for Marduk and gather in it to celebrate his mighty deeds. Enuma Elish was probably composed in the early part of the second millennium B.C.[6]

There are at least two ways to view such stories. We can focus only on our own story (whether it is the Omaha, Buddhist, or Judeo-Christian) and affirm its truth without considering other religious traditions. In other words, we can restrict ourselves to the confines of traditional religious life and commitment. Another way is to look for the human significance the stories might have for those who believe them, or even possibly for all humanity. Thus we would move beyond the confines of religions themselves. The former approach is that of traditional religious leaders, shamans, or priests, whereas the latter is that of modern scholars.

If we follow the latter approach, we note first that the stories differ in particular details yet agree somewhat in general content and form. Most of the particular differences can be ascribed to natural phenomena in the area in which the stories originate. For example, the Omaha story refers to such familiar phenomena of the American landscape as volcanoes, prairies, and wild fruit trees, while the Mesopotamian story reflects the commingling of fresh and salt water in the Persian Gulf as well as particular weather conditions in that country.

The similarities are as clear as the differences. Certain opposing pairs appear throughout the Omaha, Mesopotamian, and biblical creation stories and in numerous others as well: darkness-light, water–dry land, chaos-order, despair-joy. In most cases these similarities cannot be attributed to historical contact between cultures. Hence, it is believed that all human beings "by nature" have certain universal interests, concerns, and needs expressed through the symbols common to most of these stories. Thus, for example, both the Omaha and the biblical stories show a concern for dry land, seeds, and fruit. Obviously the focus here is on the construction of a world in which the elemental necessities of life are readily available. The stories express interest in, as well as gratitude for, a world in which these necessities are furnished in abundance. Just as we saw in the case of the biblical story, the others do not really resemble scientific descriptions and theories at all. It seems then that all ancient peoples, including the Israelites, did not have as their goal the aim of modern science, namely, the careful observation and description of the natural world, even though in Genesis and elsewhere they obviously do express views on what happened in their past and what was the case in their present. In this sense, they do give "factual" accounts; that is, they refer to items such as sun, moon, and stars. At the same time, a careful description in the modern sense is not their aim. Their accounts resemble poetic and other forms of artistic expression, which convey human interests and concerns by means of the symbolic use of natural phenomena.

This way of reading the stories casts a new light on them. They are no longer understood mainly as primitive scientific efforts to describe natural phenomena. Looking at Genesis that way, we found the stories to be very poor efforts. The same could be said of the Omaha, Babylonian, or any other creation story read in this way. Furthermore, if we regarded them only as obsolete science, we would miss the real significance that the stories had for their original users, namely, their social and religious significance. Thus the polarities mentioned above, which appear almost universally in creation stories, can be understood best as follows: human beings basically seek a comfortable, prosperous, and secure way of life; this concern is expressed through symbolic use of natural phenomena with which people have immediate experience; water and darkness are inhospitable, whereas dry land and light are hospitable. This symbolism explains the focus in the biblical story on light and land over darkness and water, and it solves all the puzzles arising from our effort to read Genesis as a scientific description instead of a symbolic expression of religious beliefs. With the polarity expressed in chaos-order the reader is invited to move directly from the natural to the personal and social worlds of the people concerned, who are fervently seeking an orderly life. And in the last polarity, despair-joy, we witness the "spiritual" expression of humankind's deepest consciousness of life in the world.

The validity of reading and understanding creation stories in this way is confirmed when we answer the following question: What was the social and institutional setting of creation stories? Although we are not certain about all cases, in many the setting is one of religious rituals. For example, the Babylonian story was told in the temple in Babylon on the fourth of the twelve days of the New Year celebration, when the story was also reenacted by religious functionaries. The intent of the myth and ritual was to guarantee Marduk's preservation of Babylonian security and prosperity. In another cultural setting, the Hindu, we notice that creation stories played a role in rituals performed when new territory was taken or buildings such as houses were erected. The themes of the creation stories were used in these instances of physical, social, and political creating and ordering. The power and sanction of the gods were brought to bear on these human actions to consecrate and legitimize them.

Can we look at the biblical creation story in a similar way? And if we can, what do we discover? We can, and we find similar results. We must first notice, however, that there are actually two stories in Genesis 1–3 that have been loosely combined to form one story. This conclusion is based on the following kind of observations. In Genesis 1:1 to 2:4a the Hebrew word that is used for god (transliterated) is Elohim, whereas in 2:4b to 3:24 the term is Yahweh Elohim (the English, Lord, actually translates a specific name of God, which then transliterated into English gives Yahweh). The precreated world in Genesis 1:2 is a watery chaos, whereas in Genesis 2:5–11 it is a desert wasteland. In Genesis 1:26–31 man and woman are created together after the whole world has been created, whereas in Genesis 2:7–24 man is created first, then woman, only after the creation of a garden in Eden. In Genesis 1:1–2:4a God is a majestic transcendent being, whereas in Genesis 2:4b–3:24 he is an anthropomorphic immanent being, who is described in homely fashion. Genesis 1:1–2:4a has a poetic liturgical pattern, whereas Genesis 2:4b–3:24 is a folksy story. The two sections differ quite radically, then, in vocabulary, specific events, order of events, theology, and style.

Modern scholars thus conclude that we have two stories in Genesis 1–3, and that these stories are the first parts of two major traditions that pervade the whole Pentateuch. Scholars also try to pinpoint the period in the history of Israel when these traditions originated and their religious message was especially pertinent. Thus the tradition of which the second story is the beginning is thought to come from the time of David and Solomon (tenth century B.C.E.). It expresses the various concerns of some thoughtful people about the social and political developments that took place under the early monarchy. Likewise, the tradition of which the first story is the beginning is thought to originate around the time of the catastrophic destruction of Judah (587 B.C.E.), when anxiety about the orderliness and meaningfulness of the world, especially the social-moral world, was especially acute. We may theorize then that the second story expresses dominantly the ambivalence some sensitive souls felt about the empire built by David and Solomon. The glory and misery of that empire, its good and its evil, were poignantly expressed in the compact tale of the first man and woman who likewise lived in a paradise, but who also tragically and mysteriously fell into sin. The first story, however, expresses the faith of the community leaders, the priests, that despite the debacle of the fall of Judah and the erosion of traditional beliefs about God and Israel, the world was still God's world of meaningful order. This is still the world created by the transcendent creator. The formal character of this story probably indicates that it was used in a ritual context.

This discussion of the two creation stories in Genesis and our earlier analysis of other creation stories indicate how modern scholars interpret the biblical stories as religious rather than scientific literature. And they make the creationists' scientific interpretation of Genesis all the more puzzling. Not only must we wonder which of these mutually exclusive stories could be the scientifically true one, but we also wonder why it is that creationists miss the religious character and import of the stories. This is indeed ironic because creationists seem to be basically religiously motivated, as we have seen above in the Creation Research Society's "statements of faith." But perhaps the creationists' insistence on reading Genesis (as well as the whole Bible) as scientifically true is itself a religious act. Why is it that creationists insist that God created the world? Is there something deeply religious at stake in the belief in a creator God? I think one scholar distinguished this when he pointed out that Genesis 1:1–2:4 is a "solemn insistence in worship that the world created by God stands fixed and secure, ordered to fulfill the divine purpose."[7] Indeed, most creator gods have as their main function the construction and preservation of a hospitable and secure world for human beings.[8] In the stories given above, this is clearly the main purpose of Wakonda and Marduk.

We can understand much about creationists from this point of view. In particular, we can understand their obsession with creation over almost all other religious themes, their insistence on the scientific truth of the biblical story, and their refusal to participate in, or even to take seriously, either modern biblical scholarship or modern study of religion. Creationists think and act thus because they find in God as creator the same kind of world order and security people have always found through their belief in a creator. The distinctive characteristic of contemporary creationists is that they interpret the religious literature of the Bible as if it were modern scientific literature in order to reinforce their religious beliefs. By insisting on the scientific accuracy of Genesis, even in the face of overwhelming contrary evidence, they hope to call upon the prestige of science to support their beliefs. Thus, ironically, it seems that these religious people try to build a foundation for their lives on scientific proof rather than on religious faith.

For Further Reading

MODERN STUDY OF RELIGION

COMSTOCK, W. RICHARD. *The Study of Religion and Primitive Religions.* New York: Harper & Row, 1972.
    A more elementary work than Eliade's. Designed for college classes.
ELIADE, MIRCEA. *The Sacred and Profane: The Nature of Religion.* New York: Harper & Row, 1959.
    The outstanding classic in modern study of religion. All of Eliade's books are important.
GILL, SAM, et al. "Creation." *Parabola* 2(1977):2.
    The whole issue deals with the phenomenon of creation in a variety of ways.
HAWLEY, JOHN S. "Krishna's Cosmic Victories." *Journal of the American Academy of Religion* 47(1979):201–21.
    Describes the function of an order-creating and -maintaining deity.
LOWRY, SHIRLEY PARK. *Familiar Mysteries: The Truth of Myth.* New York: Oxford University Press, 1982.
    Part III of this book, "The Compleat Home and the Monster at the Door," is an excellent wide-ranging discussion of the chaos/cosmos polarity.

CREATION STORIES: PRIMARY SOURCES

ELIADE, MIRCEA, ed. *Gods, Goddesses, and Myths of Creation.* New York: Harper & Row, 1974.
A good, minimal selection.

FARMER, PENELOPE. *Beginnings: Creation Myths of the World.* New York: Atheneum, 1979.
Arranged topically: Earth, man, flood, fire, death, and so on.

LONG, CHARLES H. *Alpha: The Myths of Creation.* New York: G. Braziller, 1963.
Arranged according to myth types, with introductory essays.

MACLAGAN, DAVID. *Creation Myths: Man's Introduction to the World.* London: Thames and Hudson, 1977.
Many enlightening pictures and diagrams. Focus on abstract meanings, such as "something from nothing," and "world-order and the order of worlds."

SPROUL, BARBARA C. *Primal Myths: Creating the World.* New York: Harper & Row. 1979.
The most comprehensive collection, with brief explanatory introductions.

MESOPOTAMIAN CREATION STORIES

BRANDON, S. G. F. *Creation Legends of the Ancient Near East.* London: Hodder and Stoughton, 1963.
An excellent, extended study.

GASTER, T. H. "Cosmogony," in *Interpreter's Dictionary of the Bible.* Nashville: Abingdon Press, 1962.
A brief introduction to ancient Near Eastern ideas about how the world came into being.

HEIDEL, ALEXANDER, ed. and trans. *The Babylonian Genesis,* 2nd ed. Chicago: University of Chicago Press, 1951.
The basic text for Enuma Elish, the Babylonian story to which the biblical materials are closely related.

JACOBSEN, THORKILD. "Second Millennium Metaphors, World Origins and World Order: The Creation Epic," in *The Treasures of Darkness: A History of Mesopotamian Religion.* New Haven: Yale University Press, 1976.
The best study of Enuma Elish and creation stories in general.

THE GENESIS STORIES

ANDERSON, BERNHARD W. *Creation and Chaos: The Reinterpretation of Mythical Symbolism in the Bible.* New York: Association Press, 1967.
An informative and sensitive theological interpretation of creation belief in the Bible.

ASIMOV, ISAAC. *In the Beginning.* New York: Crown Publishers, 1981.
An elementary explanation of the early chapters of Genesis with explicit reference to modern knowledge.

GROS LOUIS, KENNETH R. R. "Genesis I–II, The Garden of Eden," in *Literary Interpretation of Biblical Narratives,* edited by K. R. R. Gros Louis. Nashville: Abingdon Press, 1974.
Excellent interpretations of Genesis as religious literature by a literary critic.

SARNA, NAHUM M. *Understanding Genesis.* New York: McGraw-Hill, 1966.
Relates Genesis explicitly to modern ideas, including evolution. Makes judicious use of Enuma Elish in order to understand the role of creation stories in society.

SPEISER, E. A. *Genesis.* Garden City, N.Y.: Doubleday, 1964.
The best commentary on Genesis for getting the meaning of the text.

[11]

Christianity and Evolution

BRENT PHILIP WATERS

T HE contemporary controversy between creationism and evolution is not
actually an issue between competing scientific perspectives but is instead the latest round in a long dispute regarding biblical authority.
This chapter examines selected Christian perspectives in the controversy between evolution and scientific creationism. The discussion moves from the historical context to the issue of biblical authority, the contrasting theological
perspectives, and an argument supporting evolution as an acceptable theological paradigm for understanding the Christian doctrine of creation.

DEFINITIONS

Before proceeding with our discussion, however, we should distinguish
between two opposing schools of thought, which I shall call "liberal" and "conservative." These terms denote neither specific organizational structures within
Christianity nor any political persuasions. A liberal Christian could be a
political conservative, for example.

Liberal Christianity in the United States traces its historical roots to the
Great and Second Awakenings, the rise of denominationalism, and the social
reform and missionary movements of the nineteenth century. The split with
conservative Christianity occurred in the late nineteenth and early twentieth
centuries during the modernist-fundamentalist controversy. Intellectually, liberals tend to accept critical biblical scholarship and have been characterized by
the successive, dominant theological perspectives of the social gospel, liberalism, Neoorthodoxy, and, currently, pluralism.

Conservative Christians share a common historical heritage with their
liberal counterparts. The modernist-fundamentalist controversy caused various
denominations to splinter and intellectual interests to go in divergent directions. Generally, conservatives view biblical criticism with great skepticism and
place theological emphasis on personal piety, confessional doctrinal purity, or
formal orthodox apologetics.

HISTORICAL CONTEXT

The origins of the current controversy between evolutionism and creationism lie in the modernist-fundamentalist confrontations of the late nineteenth and early twentieth centuries. The intellectual climate in the United States after the Civil War was changing rapidly and radically. In particular, the Protestant churches were rocked by the publication of new theories in geology, biology, and critical biblical scholarship. In addition, the "new" disciplines of psychology, sociology, and comparative religion were forcing a rigorous self-evaluation. Each of these changes within the intellectual community presented a unique series of challenges that set the stage for the approaching controversy. As church historian Winthrop S. Hudson summarizes:

> Each of these several currents of thought constituted a threat to accepted understandings of the Christian faith. The psychological and sociological studies tended to reduce religion to a social phenomenon. The accounts of other religions raised questions with regard to the uniqueness of the Christian faith. Both Darwinian biology and the new biblical studies seemed to undermine the authority of the Bible. How could one reconcile Darwinism with the Genesis accounts of creation? And how could one adjust the doctrine of biblical inspiration to take account of the errors and contradictions in the biblical text that were being noted by the scholars?[1]

In response to these dilemmas, two schools of thought quickly developed. The changing intellectual climate was perceived as a threat that had to be repulsed through a reaffirmation of orthodoxy. In 1874, Charles Hodge defended the doctrine of creation and the principle of design, in opposition to natural selection. The work of Hodge and others laid the foundation for the publication between 1910 and 1913 of *The Fundamentals,* a series of twelve booklets written by a variety of conservative scholars. It was an attempt to defend orthodoxy systematically by directing attention toward divine design in creation and the inerrancy of the Bible.

Liberal or modernist Protestants accepted the findings of the new sciences and critical biblical studies, and they tried to accommodate them with their religious beliefs. This approach represented an attempt at maintaining a historical self-understanding while at the same time expressing it in intelligible and relevant forms. Henry Ward Beecher, a leading liberal, declared in 1872:

> If ministers do not make their theological systems conform to the facts as they are; if they do not recognize what men are studying, the time will not be far distant when the pulpit will be like a voice crying in the wilderness. And it will not be "Prepare the way of the Lord," either. . . .
> The providence of God is rolling forward in a spirit of investigation that Christian ministers must meet and join.[2]

For nearly five decades a cold war of polemics characterized the relationship between fundamentalists and modernists. In the 1920s, however, the conflict turned hot. According to Sidney E. Ahlstrom: "This militancy took two distinct forms: first, as an effort to prevent public schools and universities from teaching scientific theories which were deemed incompatible with traditional interpretations of the Bible; and second, as an effort to block the advance of liberal theology and modern scholarship in the churches."[3] Of these two issues, the second was primarily an internal ecclesiastical struggle that centered on a series of bitter heresy trials and attempts to control curriculum in denominational seminaries. The scope of this chapter prohibits a discussion of this "in-house" fight.

The first issue, however, relates directly to our concern. The single event that symbolized the struggle between modernism and fundamentalism was the trial of John Scopes in July of 1925. Scopes, a high school teacher, had been indicted for violating the state code that prohibited "the teaching of the Evolution Theory in all the Universities, Normals, and all other public schools of Tennessee."[4] The case itself was not unique, for Oklahoma and Florida had already passed similar legislation in 1923, and Mississippi and Arkansas would follow in 1926 and 1927, respectively. What gave the event notoriety was the confrontation of William Jennings Bryan and Clarence Darrow. Although Scopes was convicted, the cause that Bryan championed was popularly discredited in the nationwide press coverage the case received. Some individuals continued aggressive attacks against the "new heresy" and maintained a rigid defense of "revealed truth," but the debate no longer received close public scrutiny. For liberal Protestants it was a dead issue.

The upheaval of the 1920s brought attention to a variety of social and religious issues that have enjoyed renewed public interest in recent years. Although the language is different and the arguments somewhat more sophisticated, the issues are essentially the same as those addressed earlier. The central dilemma both then and now is the issue of biblical authority. Only when the differences regarding the meaning and purpose of the Bible are described, can the theological and social implications be understood. In other words, the issue of biblical authority (not scientific methodology) provides the proper context for understanding the current creationism-evolution controversy.

BIBLICAL AUTHORITY

Central to Protestant, and to a lesser extent Catholic, interpretation of Christianity is the Bible. A problem exists in how this authority is viewed and what effects it has on interpreting biblical material. This can be illustrated through a brief examination of three contrasting principles: inerrancy versus trustworthiness, ideological versus literal interpretation, and universal versus contextual meaning.

INERRANCY VERSUS TRUSTWORTHINESS

Conservative Christianity generally accepts the principle that biblical material is inerrant in every detail. A modern critic of fundamentalism explains that for fundamentalists "The bible contains no error of any kind—not only theological error, but error in any sort of historical, geographical or scientific fact, is completely absent from the Bible."[5] Or as conservative theologian Harold Lindsell maintains:

the Bible is infallible or inerrant. It communicates religious truth, not religious error. But there is more. Whatever it communicates is to be trusted and can be relied upon as being true. The Bible is not a textbook on chemistry, astronomy, philosophy, or medicine. But when it speaks on matters having to do with these or any other subjects, the Bible does not lie to us. It does not contain error of any kind. Thus, the Bible, if true in all parts, cannot possibly teach that the earth is flat, that two and two make five, or that events happened at times other than we know they did.[6]

Inerrancy is an overarching principle that not only governs theological thinking but also molds and shapes world views. In a sense, the Bible is a "given" to which all other information must conform.

In contrast, liberal Christians—for reasons like those stated in the previous chapter—reject biblical inerrancy, viewing the Bible as a "trustworthy" document. The Bible's authority rests in its ability to enlighten the human condition. The trustworthiness of the Bible does not lie in its ability to *describe* an event such as a forty-day and forty-night deluge but in its ability to *express* such principles or insights as that one ought to love one's neighbor. In this sense, biblical authority is not affected by the question of whether it agrees with scientific findings.

IDEOLOGICAL VERSUS LITERAL INTERPRETATION

When asked what separates conservative and liberal Christianity, many would reply that, unlike liberals, conservatives read the Bible literally. This is really not the case.

The principal purpose of a conservative interpretation of biblical literature is not to find its "literal" meaning but to have the Bible "be so interpreted as to avoid any admission that it contains any kind of *error*."[7] The primary goal is to maintain biblical inerrancy. I call this an "ideological interpretation." This means that the Bible is approached within a framework of prior assumptions that require a combination of literal and nonliteral interpretations depending on circumstances. The Bible is read with a bias that its statements should prove its own inerrancy.

For the liberal, the primary problem is to allow the biblical stories to stand on their own integrity and to approach them as they *literally* appear in the text. This means that poetry must be read as poetry, statements of faith cannot be seen as objective descriptions, and prescientific literature cannot be used to evaluate modern science. Instead of interpreting the Bible from ideological assumptions, one must ask questions such as, In what historical and cultural context was the author writing? and What is he or she attempting to communicate? From this perspective, it is assumed that statements in the Bible should be read literally and that they reflect differing historical contexts and literary purposes.

UNIVERSAL VERSUS CONTEXTUAL MEANING

This final contrast in perspectives on biblical authority grows out of the question of how an ancient collection of literature can have an immediate relevancy for contemporary readers.

For conservatives, the Bible was inspired by a god who transcends all historical and cultural differences. Therefore, the revealed truths are timeless and universal — they have an immediate relevancy for any time or audience. The writings of Saint Paul are addressed directly to concerns shared by residents of Corinth nearly two thousand years ago and to those residing in any town or city today.

The problems of differing times and locations are not so easily transcended by liberals. As stated earlier, liberal Christians tend to approach the biblical material as it literally appears in the text, which is reflective of distinct cultural and historical contexts. A twofold process is required to discover the relevancy of various biblical texts. First, the historical, cultural, and intellectual context must be reconstructed and the question then asked: What did the author intend to communicate? Second, once this question has been answered, roughly analogous contemporary issues must be addressed and the original intent of the author explained or interpreted through the employment of relevant forms and idioms.

Biblical authority has been, and continues to be, a central, divisive issue in Christianity. In general, conservatives believe that the Bible is inerrant, interpret it from this ideological assumption, and maintain that it has universal meaning and application for all people. In contrast, liberals tend to approach the Bible as a theologically trustworthy document, accept its literal presentation in the text, and interpret it contextually. These two differing understandings of biblical authority result in directly contrasting interpretations of the Bible.

For conservatives, such problems as textual contradictions and inaccurate reports of events require elaborate harmonization. For example, with the biblical creation stories a conservative approach assumes that the stories are true in every detail, and selected scientific evidence is employed to verify this belief. What may appear to be different accounts of creation are really separate perspectives reporting complementary details of the same event. The truth (religious and scientific) of the creation stories is as obvious today as when first written.

Liberals, however, maintain that textual problems within the Bible reflect different historical and cultural contexts. The context that produced the biblical material must be reconstructed before its religious message can be understood. The issue therefore has nothing to do with science, because scientific methodology can neither confirm nor falsify the religious content expressed in the creation stories.

THEOLOGICAL PERSPECTIVES

Regardless of how biblical authority may be understood, the need to explain and articulate religious faith remains. This is the task of theology. Within both liberal and conservative Christianity, there is a wide spectrum of thought regarding the meaning of the doctrine of creation. Since the nineteenth century advances in geology and history, especially Darwin's *Origin of Species,* theologians have been forced to reinterpret the biblical creation stories. This has been done either by reasserting a rigid orthodoxy, by accommodating the biblical material to scientific evidence, or by using evolution to explain how divine creativity takes place. These three positions can be labeled strict creationism, progressive creationism, and theistic evolution, respectively.

STRICT CREATIONISM

From this perspective, God created the universe without using any pre-existing materials. The biblical creation stories give witness to this "fact" and are supplemented by scientific evidence. Science and Scripture do not make conflicting claims but are actually in harmony, although they may employ different technical or descriptive language. The Genesis account is a "pictorial-summary" of what occurred at the origin of the universe. According to the nineteenth century theologian Augustus Strong, this means that the biblical creation stories are

> a rough sketch of the history of creation, true in all its essential features, but presented in graphic form suited to the common mind and to earlier as well as to later ages. While conveying to primitive man as accurate an idea of God's work as man was able to comprehend, the revelation was yet given in pregnant language, so that it could expand to all the ascertained results of subsequent physical research. This general correspondence of the narrative with the teachings of science, and its power to adapt itself to every advance in human knowledge, differences it from every other cosmogony current among men.[8]

This approach reflects a conservative understanding of biblical authority by selectively embracing the best current scientific evidence that "proves" the inerrancy of the bible. The procedure can be seen in Strong's apologetic explanation of the first chapter of Genesis.

Strong argued that there was a close approximation of coincidences between the biblical story and the conclusions of two leading geologists of his day, Dana and Guyot, who proposed that the earth was at first a "gaseous fluid" or "condensing nebula." According to Strong, these terms are synonymous with the biblical word "waters." The beginning of "molecular activity" resulted in the production of light independent of the sun, as described in Genesis:

> Here we have a day without a sun—a feature in the narrative quite consistent with two facts of science: first, that the nebula would naturally be self-luminous, and secondly, that the earth proper, which reached its present form before the sun, would, when it was thrown off, be itself a self-luminous and molten mass. The day was therefore continuous—day without a night.[9]

Strong explains that the separation of the waters is a reference to "primordial cosmic material" and that the appearance of land is the result of igneous action. The creation of the land is accompanied by the "creation of the vegetable kingdom," which is followed by a clearing of the vapors surrounding the earth, thus making possible the appearance of "higher animal forms" and allowing the light of the sun to become visible. Strong concludes his harmonization of science and Genesis with a description of the successive appearances of the "four grand types of the animal kingdom," mammals, and humans who possess full "moral and intellectual qualities."

Although Strong's *Systematic Theology* was written in 1896, it never-theless provides an excellent example of contemporary creationist methodol-ogy — which selects certain scientific evidence to support biblical inerrancy. To-day, the works of Dana and Guyot have simply been replaced by the "big bang" or other contemporary theories. Regardless of what claims to objectivity might be made by scientific creationists, their methodology does not lead to open discovery but attempts to prove prior assumptions interpreted from biblical material. As a result, creationists tend to distort or misapply scientific theories, as exemplified in their use of entropy to rebut evolutionists' argu-ments (see Chapter 9). On the other hand, scientific evidence that challenges the creationist view is excused or overlooked. Strict creationism, or the so-called scientific creationism, is simply the selective use of available data to support a particular theological perspective and understanding of biblical authority.

PROGRESSIVE CREATIONISM

Progressive creationism shares with strict creationism a basically conserv-ative understanding of biblical authority, with some liberal overtones. The Bi-ble is essentially correct in all statements but does not provide criteria for ac-cepting or rejecting modern scientific theories. The primary difference is a pro-gressive creationist's understanding and use of scientific evidence. Not only are the arguments more sophisticated but the influence of scientific theories is much more apparent in the theological methodology and conclusions of the progressive creationist.

What is striking about this position is the unwavering belief in the Bible's essential correctness and the struggle this creates in attempts to harmonize science and Scripture. It is a classic example of an ideological understanding of the Bible. Since the biblical material is true, it will be interpreted to corres-pond with widely accepted scientific information, while scientific conclusions that appear to be supportive will be accepted and those that do not will be rejected.

Central to this understanding of the doctrine of creation is the absolute need to harmonize the discoveries of modern science with the Bible. The im-portance of this endeavor lies in the belief that the God of biblical revelation is the same God who created the natural world:

> If we believe that the God of creation is the God of redemption, and that the God of redemption is the God of creation, then we are committed to some very positive theory of harmonization between science and evangel-icalism. God cannot contradict His speech in Nature by His speech in Scripture. If the Author of Nature and Scripture are the same God, then the two books of God must eventually recite the same story.[10]

To facilitate this harmonization, as conservative scholar Bernard Ramm explains, one must recognize that the language of the Bible is popular, pre-scientific, and nonpostulational. This provides a great deal of latitude. Thus

Ramm is able to reject various creationist interpretations and present his own concept of progressive creationism, which is based on general principles, rather than on a direct correspondence between the sequence of creation in Genesis 1 and available scientific evidence. He notes that since the language of the Bible is popular rather than technical, the term "day" does *not* refer to a twenty-four-hour period. The biblical language represents categorical statements that lump common concerns together, which are topically rather than chronologically presented. In other words, it is like saying that "God created the universe" and then listing examples as they come to mind without regard to logical order. Ramm offers this illustration:

> A carpenter can tell his child that he made a house — the roof, walls, floors, and basement. The child realizes that his father made the house even though the father gave a topical order, not a chronological order. The creation record is part topical and part chronological to convey to man: (i) some sense of the order in creation; (ii) that God made everything, so nothing may be worshipped. Man as the last in order is highest in importance, and for that truth the order is necessary.[11]

Creationism and science, therefore, are not in conflict but are different descriptions of the same event. Thus the Bible, if interpreted correctly, is verified by selective scientific evidence that corresponds to the biblical account.

Finally, Ramm briefly explores the role of evolution in the scheme. He rejects any form of "atheistic evolution" but looks favorably upon "a spiritualistic interpretation of evolution." This means that a *qualified* acceptance of evolution is permissible if it is viewed as a secondary cause that was preceded by an original or actual cause. These qualifications include the stipulations that all theorizing about evolution must "reckon with energy in design in nature" and that the transcendental nature of humanity is a result of divine intervention rather than a product of natural processes. Although evolution has guided the natural development of the world since its inception, all ultimate origins are the result of divine acts.

THEISTIC EVOLUTION

Theistic evolution grew out of the attempts of early liberals to explain the doctrine of creation by means of evolutionary principles. It was a position held by many influential Protestants and Catholics in the late nineteenth and early twentieth centuries. It differs from strict and progressive creationism in maintaining that within limits evolutionary principles do not run counter to the biblical accounts of creation. Mutations and adaptations do occur within species, and such changes do not refute a biblical understanding of creation. A divine being has initiated and at times intervenes in natural processes. The evolutionary process is simply the method God has chosen to direct the universe. As J. C. Jones writes:

> A modified form, however, of the doctrine of evolution is possible — that which I have named supernaturalistic. This admits of immediate Divine in-

tervention at certain critical moments in the creation and development of life, by which God introduces new staple into the loom, new material into the machine. But these interventions should not be viewed as so numerous as to crowd the universe with miracles. The laws of nature are allowed to work out their own results for millions of years—results seen in innumerable varieties and improvements of species.[12]

Theistic evolution differentiates itself from natural evolution with respect to humans. Although natural change and adaptation are accepted, the origin of humanity is an exception. Views range from a special creation in an isolated act of God to the belief that if humans did develop from animal-like ancestors, the moment in which they became human reflects a divine intervention. In other words, evolution characterizes the natural world, but humans transcend its effects and limitations.

Although evolution may be the way God works in nature, theistic evolutionists share in common with creationists the belief that origins and destiny lie outside natural explanations. It is assumed that matter, or the physical universe, does not exist eternally, and this implies a distinct beginning and end. A further implication is that there is a purpose or teleological element to the universe—God created, governs, and directs creation toward an eventual goal or destiny. Strictly speaking, a theistic evolutionist considers evolution to be a natural interim between the supernatural events of original creation and ultimate conclusion. In theological terms, evolution is the concern of the doctrine of divine providence rather than the doctrine of creation.

EVOLUTIONARY CREATIVITY

Evolutionary creativity differs from the previous views of evolution primarily in its understanding of God. Although the perspectives of strict creationism, progressive creationism, and theistic evolution disagree on the means that were divinely employed to bring about creation, they share a common doctrinal understanding of the nature of God. Traditionally, God has been defined as an independent being, wholly removed, and standing over and against the natural order. An "absolute gulf,"[13] as Emil Brunner explains, separates the created from the supernatural creator. God created, out of nothing, the natural order and directs its progress to an ultimate conclusion that has always existed within the "mind" of God. This is a static understanding of deity.

In opposition to this traditional explanation, there developed a radically dynamic understanding of God and the meaning of the doctrine of creation. Borrowing from the philosophy of Alfred North Whitehead, process theologians see God as being intrinsically related to and affected by the natural order. There is no absolute gulf between the creator and created because such a distinction is both arbitrary and meaningless. The natural order is a moment-by-moment process that re-creates or renews itself. The process takes place within that which we name God. Rather than being the originator of a sudden and purposeful act of creation, God attempts to direct or lure the natural creative process. As John B. Cobb, Jr., explains: "That means that the Creator-Lord of history is not the all-determinative cause of the course of natural and historical events, but a lover of the world who calls it ever beyond what it has attained by affirming life, novelty, consciousness, and freedom again and again."[14] The issues of origin and destiny can never be precisely known because they lie beyond the boundaries of the process itself. "The beginning of reality is ultimately unexplainable. The only explanation available to us comes from looking at the nature of process as a whole."[15]

A process-theological interpretation of the doctrine of creation represents a liberal understanding of biblical authority—the biblical creation stories are viewed properly as mythic literature. The attempts by ancient people to explain their origin provide no scientific description but are expressive of existential concerns. The task of the theologian is to identify this concern and articulate it in a relevant and meaningful manner. A supernatural creation is rejected, but divine participation in the natural process is maintained. God is seen as molding or directing existing matter from chaotic to useful forms. A changing or evolutionary process is therefore implied within the biblical creation stories themselves.

Evolution takes on great importance because it is used as a paradigmatic device to articulate a theological understanding of creation or the creative process. A paradigm, in this sense, is used as a device to interpret data gained through experiences and perceptions. A theological explanation of God and creation must be representative (rather than exceptional) of our understanding of the natural order since the divine is intrinsically related or linked to it. The importance of the evolutionary paradigm is that it interprets creation as a process of continual "becoming" rather than an isolated event of supernatural origin. As Pierre Teilhard de Chardin explains: "Is evolution a theory, a system or a hypothesis? It is much more: it is a general condition to which all theories, all hypotheses, all systems must bow and which they must satisfy henceforward if they are to be thinkable and true. Evolution is a light illuminating all facts, a curve that all lines must follow."[16]

The use of evolution as a paradigmatic device results in an understanding of creation far different from those previously examined. The best known and perhaps most ambitious advocate of this perspective was the paleontologist and Jesuit priest Pierre Teilhard de Chardin.

TEILHARD'S VISION

Within Teilhard's elaborate vision, the evolutionary process may be divided into four general phases: prelife, life, thought, and survival. Throughout each phase run two interconnected and continuous threads. First is the principle of "cosmogenesis"—the viewing of the "becoming" of the cosmos as a whole. Second is the process of "noogenesis"—the gradual development of mind and mental capability.

Within the prelife phase, Teilhard maintains, the "stuff of the universe" was composed of elemental matter, total matter, and the appearance of evolutionary processes. When pushed back to the farthest point, everything within the universe has a common natural origin: "To push anything back into the past is equivalent to reducing it to its simplest elements. Traced as far as possible in the direction of their origins, the last fibers of the human aggregate are lost to view and are merged in our eyes with the very stuff of the universe."[17] The potential for life existed within the inanimate stuff of the universe, and its appearance must be sought here rather than in a unique supernatural intervention. The emergence of life is the result of a "tangential energy" that links all elements together and a "radial energy" that moves toward an ever-increasing complexity and centricity. Teilhard insists that the potential creativity of life *always* existed in the "within of things" and cannot be explained by appeal to such an external source as an isolated divine being.

The advent of simple life forms represents the first of several quantum leaps in evolutionary creativity. Teilhard explains that the initial leap was the result of an "external revolution": "From an external point of view, which is the ordinary biological one, the essential originality of the cell seems to have been the discovery of a new method of agglomerating a larger amount of matter in a single unit."[18] In addition to this external revolution, an "internal revolution" took place in which a primitive consciousness was already in the early stages of development within simple cells. From this beginning, life rapidly expanded through a series of movements: reproduction, multiplication, renovation, congregation, association, and controlled additivity. The primary theological point that Teilhard makes is that creation must be viewed as an ongoing natural process of creativity that is not dependent upon external sources to provide unexpected "bursts of energy" for the appearance of life. Human life must see its interconnectedness with all matter, since its origins lie within a common natural process rather than being an isolated event of unnatural or supernatural origination.

The next quantum evolutionary jump is the development of thought or the "humanization of the individual." By this, Teilhard means the ability to develop self-awareness. From the development of self-awareness came the "deployment of the noosphere," an enlarging sphere of mental capacity promoting creativity, and the radical recognition of the cause and effect of choices and decisions. Teilhard's theological goal here is to provide a religious vision that adequately explains a creative process and development *and* actively works to keep the evolutionary possibilities open for the future. Ultimately, Teilhard suggests that creation cannot be understood as *sudden appearances* but instead as a metamorphosis of existing material.

Finally, Teilhard believes that we are on the brink of another evolutionary leap, that of survival and development of the "ultimate earth." Since the primary emphasis of this chapter is upon origins rather than possible destinies, most of what Teilhard says at this point does not fall within our scope. One point, however, should be mentioned. The creative evolutionary process is heading toward "omega point" at which the personal will be synthesized with the postpersonal. Teilhard implies that there is a purpose (omega point) toward which the creative evolutionary process is headed. Unlike traditional teleological goals, however, the purpose of the universe was not predestined or preordained in the beginning by divine authority or mandate. Instead, omega point is a goal that is itself enlarging and changing, depending upon what direction the evolutionary process may take. Omega has been involved in, and will grow out of, the creative evolutionary process.

Contrary to the claim of some creationists that, without appealing to the Bible, scientific arguments can be offered that imply supernatural acts of creation, their doctrine presupposes a highly conservative interpretation of the biblical creation stories. Traditionally, conservatives have interpreted the Bible from an inerrant, ideological, and universal perspective. As already discussed, this poses some peculiar problems in interpreting the biblical creation stories. Furthermore, it transforms the Bible into a filter through which scientific data must pass. The acceptability of scientific evidence is judged on whether it supports or challenges a conservative interpretation of the creation story. The result of this approach is a closed rather than open system of inquiry.

Historically, modern creationism did not grow out of a scientific discovery, but from a refusal to accept new scientific methodologies and a reaffirmation of a rigid orthodoxy. The dispute, therefore, is over the meaning of biblical authority. It is unfortunate that the issue has spilled out from the church into the public sector, as we have seen in recent court cases and legislative debates. As the various theological perspectives indicate, how individuals perceive scientific data when applied to different views of biblical authority results in contrasting cosmologies. This is the realm of theology, not science. Scientific ideas can *inform* theological constructs, but the process cannot be reversed.

The issues of biblical authority will probably never be satisfactorily resolved among Christians. In this ongoing dispute, creationists should not present their viewpoint as an exclusive one that represents the only Christian perspective on the issue. It is not only merely one among many but has long ago been discredited. The implication that evolution is "atheistic" and creationism is "Christian" is a false dichotomy. As a wide spectrum of theological viewpoints illustrate, legitimate faith does not depend on one's acceptance or rejection of evolution.

The domains of religion and science are separate and need to remain that way. Although science may inform theology, theology may not properly govern scientific discovery. Despite the best efforts of the church to defame Copernicus, the earth has not moved to the center of the universe. Scientific evidence cannot be altered by religious fiat. The current controversy is the result of addressing improper questions to the biblical material. Science asks how and when; religion asks why and what does it mean. The biblical creation stories offer a tentative answer to the latter set of questions, but they are ill equipped to provide an intelligent response to the former.

For Further Reading

BARBOUR, IAN G. *Issues in Science and Religion*. New York: Harper & Row, 1966.
    Both a scientist and theologian, Barbour examines a wide variety of issues including evolution.
BARBOUR, IAN G. *Myths, Models, and Paradigms*. New York: Harper & Row, 1974.
    This text compares and contrasts religious and scientific methodologies.

BARR, JAMES. *Fundamentalism*. Philadelphia: Westminster Press, 1977.
  Barr, Oriel Professor of the Interpretation of Holy Scripture at Oxford, presents a comprehensive and critical review of conservative theology and use of the Bible.
COBB, JOHN B., JR. *God and the World*. Philadelphia: Westminster Press, 1969.
  This leading process theologian argues for an alternative understanding of God different from that of traditional orthodoxy.
COBB, JOHN B., JR., AND GRIFFIN, DAVID RAY. *Process Theology*. Philadelphia: Westminster Press, 1976.
  This book introduces and describes basic ideas in process theology.
GILKEY, LANGDON. *Religion and the Scientific Future*. New York: Harper & Row, 1970.
  This liberal theologian argues at length that religion and science play complementary rather than adversary roles.
MELLERT, ROBERT B. *What Is Process Theology?* New York: Paulist Press, 1975.
  Mellert provides an introductory text on the philosophy of Alfred North Whitehead and its application to theology.
TEILHARD, DE CHARDIN, PIERRE. *The Phenomenon of Man*. New York: Harper & Row, 1959.
  This book provides a comprehensive interpretation of the doctrine of creation from an evolutionary point of view.

[12]

A Liberal Manifesto

HENRY A. CAMPBELL

THE liberal church in the United States stands at a critical juncture. Typical of this is Union Church, a large suburban church in Illinois, which, after a decade of declining membership, formed a long-range planning committee to investigate the situation. After several months' work the committee came face to face with the fundamental questions: Is the theology we espouse and seek to live by adequate to sustain us through the 1980s, at a time when cultural changes are taking place at an unprecedented pace? Does the basic message we have been teaching and preaching over the years need new, modernized interpretations and new focusing? The answer was yes.

The liberal church accepts as a perennial task the need for a reexamination of the theological expression of its faith and its particular forms such as worship, education, and service. It seeks to recast its faith in terms relevant to the times and the changing needs of people. This does not mean a rebuilding from the ground up—rewriting the faith, so to speak. For we have an institution rooted deeply in history and, above all, a book that is the continuing source and summary of this faith. But we do have to rethink and reinterpret this faith and try to relate it to the culture in which we find ourselves; that is a mandate of the liberal position. Otherwise, our faith will become a formality, separate from the real pulse of life, and eventually will decay from within.

The liberal churches in this country today face a difficult and exceedingly challenging, even exciting, task. The religious pendulum has swung sharply toward the fundamentalist Christian viewpoint. During the past ten years, the various streams of conservative Christianity have been growing rapidly, whereas the liberal churches have been losing ground. They are no longer the "mainline" churches representing a kind of unofficial establishment, as they had from the beginning of our nation. As the insecurities and uncertainties of modern life multiply, the "born-again" Christian movement in its claim to offer clear and final absolutes has become something of a tidal wave. This phenomenon is currently aided and abetted by a galaxy of television evangelists and other proponents of the movement who have deeply penetrated into the mass media. To complicate the situation further, this same period has been marked by a growing number of defections from the church altogether—of those who are simply uninterested in religion or think that it is strictly a private affair.

Between these two extremes—fundamentalist religion on one hand and secularist disengagement on the other—the liberal church occupies a dangerous but hopeful position, a middle ground that holds the potential for mediation between the fractured extremes of American society. The challenge, as one spokesman for liberalism has observed, is to "discover how to walk the tightrope between Christ and culture without falling into the abyss of contemporary culture, without losing sight of the distinctive vision of the Christian message."[1] In other words, the liberal church (and this has nothing to do with political labels) has an opportunity to speak to and reach millions who are not swept up in this born-again, fundamentalist tide, but who still yearn for a meaningful faith and a Christian community and liturgy to help structure and reinforce that faith. Even conservative Christians, when they come to realize the full implications of a narrow anti-intellectualism contradicted by the main currents of contemporary thought, may find a more comfortable home in one of the liberal denominations.

As yet, however, the liberal churches have not responded to this opportunity, have not set forth in clear and unmistakable terms the liberal alternative. But the time is ripe for the reassertion of the liberal expression of Christianity, with its appeal to the head as well as the heart. In the midst of our present disarray we need to rediscover our identity and share our message with a deep sense of conviction and as much excitement and vitality as seem to be evidenced in the conservative churches.

Pluralism of expression and practice in religious matters has always been a hallmark of the American way, and it should be preserved if our religious traditions are to meet the various needs and temperaments of people. Threatening this rich variety is a coalition of right-wing religious movements that wishes to impose by law, censorship, and intimidation a rigid conformity and a narrow set of values that violate the American spirit. Liberal Christianity is in the forefront of those resisting this movement, hoping to reaffirm the right to share its own distinctive values and traditions with those of like mind. But it does not pretend to know the ultimate truth or final answers to moral and ethical questions. It offers one way to truth, or rather, the framework for what it believes is an ongoing search for truth about God and the meaning of life.

Liberal Christianity offers an alternative to fundamentalist religion in four principal areas: the first two involve its relationship with the broad concerns of society, the outward looking stance of liberalism; the second two arise from within the household of faith. To understand these four benchmarks, we should briefly trace their history to determine the source of the liberal religious spirit. Clearly, the philosophical source of liberal theology and its various ramifications is the work of the enormously influential eighteenth-century German philosopher, Immanuel Kant.

Kant, who stood within the pietistic Christian tradition, assumed as a major scholarly task a reinterpretation of religious faith that would make it credible in the face of attacks by ascendent skeptics, typified by David Hume. Kant saw all forms of dogmatism as the enemy—whether of religion, science, or any system of thought. He believed that to combat dogmatism, religious beliefs had to be reinterpreted in the light of reason and knowledge. Knowledge, he argued, consists of two realms, the noumenal and the phenomenal. The noumenal is the experiential world of feeling and intuition, which includes faith, love, beauty, and the moral will. The kind of knowledge or truth gained in this realm is properly the domain of religion. The phenomenal refers to the world of objective, observable data, which is, of course, the proper domain of science. Since Kant's deep faith was accompanied by an intense interest in all areas of scientific development, the great thrust of his philosophic enterprise was the reconciliation of these two basic areas of human experience — science and religion.

Interestingly, from the standpoint of the modern situation and the gathering power of the religious right, Kant also had to struggle with the combined forces of religious and political repression and had to make an insistent claim on behalf of free inquiry and intellectual freedom. The direct line from Kant to our present situation will be evident in other ways as the chapter progresses.

Returning to the benchmarks of liberal Christianity, the first is free intellectual inquiry. The liberal church cherishes diversity of theological viewpoint and the exercise of one's mind and conscience in wrestling with religious questions. The American colonies had their experiment with repressive, conformist religion in New England Puritanism, which for a brief period managed to stifle all dissent and divergent views. In the long run, the more tolerant spirit of the Pilgrims prevailed, a spirit effectively summarized by the Pilgrim father John Robinson: "God has yet more light to break forth from his Holy Word." Gradually, the grip of strict Calvinism was loosened to allow local churches to establish their own convenants for membership without having to conform to a rigid test of creedal orthodoxy.

In time there emerged a type of church that was theologically permissive and encouraged and supported theological exploration, innovation, and the continual testing of faith in the light of new knowledge. It is this heritage that the liberal church seeks to reaffirm today.

Nowhere is the difference between fundamentalism and liberalism more evident than in their approaches to the creation stories in Genesis. Because fundamentalists are totally committed to a literal interpretation of the Bible (even allowing for variations in the degree of literalism), they insist that the Genesis accounts be treated as historically and scientifically true, and they would bend science to protect that truth. Liberals argue that the accounts in Genesis are not science and were not intended to be the result of scientific observation, for science as we know it was nonexistent at the time that these accounts were written.

Biblical scholarship since the midnineteenth century has been devoted to understanding the biblical writer's background and intent, the historical circumstances of the writing, and interrelationships with other biblical writings and nonbiblical sources. Admittedly, the critical-historical approach to the Bible cannot be substituted for the personal encounter based on faith. But by accepting these historical presuppositions, we free ourselves of the need to reconcile the ancient stories with scientific phenomena and thus can accept them at the religious or noumenal level, where they do indeed reveal truths about universal human experience. Freed from literalism, the Bible can then speak to us in its timeless wisdom and we can lay to rest the controversies, such as creationism versus evolutionism, that have consumed so much needless religious energy.

Liberalism recognizes that human knowledge will continue to be enlarged as long as humans inhabit this earth and that the religious approach is one valid way of expanding human awareness. Religious knowledge grows out of our inner experience as we seek to put together the puzzle of human life. And it is the wisdom of humility that looks beyond the limits of human understanding to the undecipherable mystery and source of it all, which we call God. The Bible is seen as the preeminent source of that different kind of unfolding knowledge, not as a narrow funnel through which all secular or scientific truth must pass.

Because of its openness to free inquiry, liberalism is sensitive to cultural traditions and the arts. Liberal Christians are alert and responsive to the amazing variety of forms in which essentially religious questions are raised. They look to all kinds of cultural expression—the novel, drama, dance, and film—for the real face of modern life and therein find the existential questions that lurk in the heart of every person. To speak intelligently and meaningfully to secular as well as religious men and women, liberalism takes seriously artistic expressions of the human predicament.

It follows, then, from this benchmark of liberal Christianity—the unfettered mind—that all attempts to subvert the quest for truth and the free flowering of human creativity must be strongly resisted. All attempts to shackle the modern mind and spirit to religious dogma must be met head on. This includes attempts to impose creationism on the schools; censorship of books in libraries, textbooks in classrooms, and movies in the marketplace; and uniform moral standards on all of society.

A second benchmark of the liberal approach is the commitment to social justice, which is poorly understood by many members of various churches, and even less by born-again Christians. People have often asked: "Well, why does the church have to get involved in race relations, or in fairness in broadcasting, or in issues of war and peace; why do church leaders and ministers participate in marches and become involved in all these tensions of our times? Why can't the church just stick to the Bible? Why can't the church simply try to build Christian character and responsibility and leave it up to individuals to work for a better world?"

The church has struggled with those questions since colonial days, and the answer has always been that to be faithful to the opening commission of Christ we must give more than lip service to social needs. Remember, Jesus began his ministry with these words: "The Spirit of the Lord is upon me . . . to preach good news to the poor . . . to proclaim release to the captives . . . to set at liberty those who are oppressed" (Luke 4:18). In other words, we are called to witness to a God who is not just interested in our worship and in making a church that is loyal to him, but to a God who is interested in the bodies, the minds, and the total welfare of all his people.

It is tempting to keep religion within the four walls of the sanctuary, to resist the promptings of the spirit to become involved with these larger issues, these concerns of people. Yet the greater pull of the spirit demands faithfulness to the God whom the Bible portrays as the champion of all kinds of human causes. To maintain the integrity of our faith, we will certainly have to step into the secular arena on occasion to witness to our beliefs and our human compassion. This precious part of the liberal Christian heritage is a reflection of the general spirit of the whole Bible as well as particular passages in it. Conservative religion, on the other hand, has always shut itself off from these larger social or human concerns, and instead has emphasized individual and local acts of charity. Ironically, the most visible expression of this tradition today has abandoned its privatism and has entered the political arena with a vengeance, adopting stances that seem diametrically opposed to the human and humane spirit of Jesus.

Consider now the third area in which our liberal tradition has been strong but in which the contemporary voices of that tradition have been weak. In its engagement with social issues the liberal church has often seemed to substitute the passing of resolutions at synods and councils and various forms of social activism for personal faith based upon the Bible and the disciplines that grow out of that faith.

It must be recalled that the greatness of our heritage lies in an impassioned combining of personal piety and social involvement in commitment to Christ and compassion for society. This combination is exemplified in Washington Gladden and Walter Rauschenbusch, both founders and leading exponents of the nineteenth-century Social Gospel movement. As pastor of an influential church in Columbus, Ohio, Gladden was a fearless social prophet, yet his sermons radiated a kind of inner mystical joy and discipline, as revealed in his great hymn, "O Master Let Me Walk with Thee." He was also a foremost reconciler of the new science with Christianity.

To those who miss this note in the liberal churches, we say, yes, the church is still in the business of helping to nurture personal faith. The Bible is held no less highly by liberal Christians than by those who insist on a literal interpretation. It is for us the wellspring of our faith, forever fresh and fascinating as it casts light on our ongoing spiritual journeys.

Loyalty to the teachings of Jesus and the character of his whole life is central to our tradition. But the liberal understanding of Christ is quite different from the fundamentalist view, as is evident from our radically different positions on social questions. Jesus' commandments are difficult to obey, for they are at odds with some of our natural human inclinations, but they are not laws cast in concrete. They unfailingly reveal a warmth and a compassion, a breadth of human sympathy and understanding that continually catches us up short. They reveal a man for others, unconcerned with wealth and success and making it big. They reveal a man not narrowly chauvinistic or bound by blind patriotism to a particular nation but one whose loyalties lay in a transcendent vision of a God who stands above all human boundaries and formulations. The commandments reveal a flexible attitude regarding the law and an insistence that it be subservient to human need—as demonstrated when Jesus allowed his hungry disciples to pick grain to eat on the sabbath, or when he broke the sabbath to heal the sick, or when he crossed the traditionally wide gulf separating women from men.

It is in the human example and the transcendent understanding of Jesus that we find inspiration and encouragement and hope to live courageously on behalf of justice in an unjust society, to witness for peace in a warlike world, and at the same time to cultivate disciplines of mind and heart that will enable us to walk as free persons with serenity and peace.

The final benchmark of the liberal church is its distinctive practice of Christian community. In a sense, the vitality of any religion lies in the communal spirit, the life of fellowship engendered there. The liberal church brings its own unique emphasis to this fellowship, which is at the heart of all religions. Since it is not organized around a doctrinal consensus, it must find other ways to foster a sense of connectedness. And in these days, it cannot depend on the laissez-faire approach "whoever seeks us out, well and good." This detached attitude must give way to genuine caring and nurturing and a recognition that we are all pilgrims needing the warmth and companionship of each other.

In this way the liberal church will see itself as a communal sanctuary — a way station where human beings can temporarily step aside from the drumbeat of conformities and expectations, where the coercive forces that bend us out of shape may for the time being be laid aside. It will be a place where we may stop and visit for a while, like the desert Bedouins gathered around the campfire, a place where we can tell our stories, share our journey — a place of radical openness where we will not be influenced by hidden agenda or motives or be manipulated toward some predetermined end.

It will be a place where we can share our values, explore our differences, and find support in our commitments. Every individual, especially in these times of narcissistic concerns, needs to find a cause, a goal, or something greater than one's self. The church lifts us out of ourselves and helps us to find these larger aims to which we may give ourselves if our full humanity is to be realized. Without this caring fellowship, this communal base, all our noble preachments and intellectual freedom count for nothing.

The key to this fellowship in a liberal church is openness, acceptance, and respect for personal differences and idiosyncrasies, life-styles, and theological viewpoints. This rich mixture gives tang and spice to our life together and makes real growth possible.

In summary, liberalism is more than simply a bulwark against a resurgence of narrow, sectarian attempts to coerce upon us all a particular ideology and ethic. It provides a theological orientation of openness to the mystery of God's grace and his amazing graciousness in the midst of human life; it gives a corresponding welcome to the rich diversity of thought and influence that surrounds us all.

The liberal church has a long heritage of open-mindedness to the cause of truth, from whatever sources, and staunchly defends and encourages free inquiry, from within science and religion, in all creative expressions of the human spirit. It also has a long history of commitment to social justice as one necessary response to the biblical prophetic tradition. The social vision of liberalism is shaped by the humanistic and compassionate Christ of the Gospels, who is revealed throughout these works as one seeking out the poor, the forsaken, and those victims of society's narrow righteousness. Thus the liberal church is no less rooted in faith in Christ than the fundamentalist, but it understands Christ as a liberating rather than a restricting force. And finally, it seeks to shape a community of faith that is tolerant and accepting of great diversities of idea and temperament: "The beauty of liberal Christianity is that one need not divorce the mind from the body, the emotions from the intellect, the passions from religion, or culture from piety."[2]

FOR FURTHER READING

BURTT, EDWIN A. *Types of Religious Philosophy*. New York and London: Harper Brothers, 1939.
   Traces the philosophic roots of every major modern religious and theological position.
DEWOLF, L. HAROLD. *The Case for Theology in Liberal Perspective*. Philadelphia: Westminster Press, 1959.
   A competent examination of the various tenets of liberalism.
MILLER, DONALD E. *The Case for Liberal Christianity*. San Francisco: Harper & Row, 1981.
   A recent reexamination of liberalism by a sociologist of religion, with reference to recent social and cultural trends.
ROBINSON, JOHN. *Honest to God*. Philadelphia: Westminster Press, 1963.
   A popular exposition of various categories of the new theology by an English theologian and bishop.
SHAPLEY, HARLOW, ed. *Science Ponders Religion*. New York: Appleton-Century-Crofts, 1960.
   A collection of essays representing various viewpoints and fields of science.
TILLICH, PAUL. *Dynamics of Faith*. New York: Harper and Brothers, 1957.
   A summary statement on the nature of faith by one of the leading Protestant theologians of the twentieth century, the theologian to whom liberals most readily turn.
WIEMAN, HENRY N. *The Wrestle of Religion with Truth*. New York: Macmillan, 1927.
   An early "process" theologian greatly concerned with the interrelationships of science, philosophy, and religion.

IV

Creationism in Public Education

[13]

Religion in the Schools:
A Historical and Legal Perspective

DONALD E. BOLES

THE current controversy over evolution and creationism in the schools of Arkansas and other states of the union is not unique. Indeed, as recently as 1968 the United States Supreme Court struck down a somewhat related statute in Arkansas in the case of *Epperson* v. *Arkansas*.[1] The Court found violative of the freedom of religion clause of the First Amendment an Arkansas law that prohibited the teaching in state-supported schools and universities of the "theory or doctrine that mankind ascended or descended from a lower order of animals." Justice Abe Fortas, speaking for a unanimous Court, recognized that the Arkansas statute was part of the "fundamentalist" syndrome that had earlier given rise to the celebrated case of *Scopes* v. *Tennessee* in which a similar "monkey law" was upheld by the Tennessee Supreme Court.[2] In *Epperson,* however, the Court found that the "law" conflicted with the constitutional prohibition of state laws respecting an establishment of religion or prohibiting the free exercise thereof. The state law, the Court explained, selects from the body of knowledge a narrow segment that it prohibits solely because it is deemed to conflict with a particular interpretation of the book of Genesis. In a sentence that could well govern our thinking concerning the law of religion in public education, the Court reminds us that as early as 1872 it had said, "the law knows no heresy and is committed to the support of no dogma, the establishment of no sect."

In beginning our historical and legal survey of problems involving religion in the schools, we should keep in mind that controversies over the relationship between church and state have shaken society to the roots from the earliest moments of recorded history. Unlike most of the nations in the Western world, the United States, although troubled to some extent by these disputes, has escaped the extremes that have plagued other nations. This has been traceable in large measure to the First Amendment to the United States Constitution and to the interpretation of it by the United States Supreme Court.

Through its founders, the United States has made two great contributions to western civilization—the concept of the separation of church and state, and the secular public school system. True, the integrity of these two principles is often compromised while their virtues are being extolled, and they are frequently honored more in the breach than in the fulfillment. But it cannot be denied that they are important planks in the platform of American political philosophy.

Historically, those who have opposed the theory of separation of church and state have centered their attacks on the public school system. This method of attack is still being used, as is apparent to all who keep abreast of the news and follow the activities of the Supreme Court even superficially.

The two propositions basic to our public school system are that public funds shall not be granted to sectarian schools and that sectarian instruction shall not be given in the public schools. Most Americans concur with these principles in theory but cannot always agree as to how they are to be practiced. This disagreement among those who sincerely believe in the separation of church and state has allowed its opponents to take advantage of the resulting confusion. They have been able to siphon off public funds for parochial schools and to introduce into the public schools reading, prayers, and other religiously oriented practices and exercises.

Before proceeding with the discussion, we should remember several basic legal concepts. First, the United States Constitution is what the Supreme Court says it is in the field of religion as in other fields. Furthermore, in dealing with religion the United States Constitution is unique. Not only is the freedom of religion guaranteed but the governmental establishment of religion is prohibited. The Supreme Court has insisted upon viewing these two clauses as separate and distinct entities. This view in turn has given rise in recent years to the recognition that on occasion these two clauses may indeed conflict with each other, particularly in areas of practice affecting the public school system.

The important implication of the distinction between the two clauses on religion is that no government coercion is required for a government-sponsored program to violate the "establishment clause." Compare this situation with that in Burma or England, which assure freedom of religion, but at the same time have an established church. In the United States, on the other hand, the courts in some instances have to choose which of these two clauses should prevail in a given situation. As is so often the case when the courts must choose between one or another of the guarantees of the Bill of Rights, there are obvious reasons for discontent and disagreement among significant numbers in the nation.

It has been suggested that two fortunate historical coincidences are responsible for the tradition of religious freedom and disestablished religion in the United States. First, this country was colonized during a period of religious revolution and counterrevolution. Second, the United States Constitution was adopted during a period of intellectual rationalism and skeptical enlightenment. As Leo Pfeffer has explained, "Neither by itself would suffice, but together they resulted in the First Amendment."[3] These two observations should be kept in mind as we examine the historical and legal development of concepts concerning religion in the schools.

PURITAN BACKGROUND

The basic pattern for the public school system of the United States was formulated in the Puritan commonwealth of colonial America. To escape the unceasing and many-sided religious antagonisms common in Europe, dissenters of various religious sects sought refuge in the New World. But once in America, these groups created established churches of their own in most of the colonies. This meant that the established settlers were not at all ready or willing to accord religious freedom to others who settled in their area and who might hold dissimilar views.

The Puritans of New England during the early seventeenth century provide the best example of the single-establishment principle. Not only was suffrage limited to church members (the clergy decided who qualified for membership) but the most vigorous methods were used to discourage dissenters from making their views known. The predominant attitudes in the New England colonies were summed up by Nathaniel Ward in *The Simple Cobler of Aggawam:*

> I dare take upon me to be the Herald of New England. So far as to proclaim to the world in the name of our colony that all Familists, Antinomians, Anabaptists and other Enthusiasts shall have free liberty to keep away from us and as such as will come be gone as fast as they can, the sooner the better. . . . He that is willing to tolerate any religion . . . either doubts his own, or is not sincere in it.[4]

It is obvious from this passage that there could be no free exercise of religion in colonial New England.

The militant Calvinism of the Puritans was accompanied by a great interest in education, partly because most of the founders of the Puritan commonwealth were educated men who were more conscious of the benefits of education than was the average person. More important, Calvinist theology demanded that every individual read, evaluate, and interpret the Bible in addition to Calvin's works. To do so, of course, required that every church member be able to read. (And, it might be added, to read Calvin one must read well.)

Furthermore, frequently overlooked is that the Puritans were also humanists. Particularly during the early stages of their settlement, they were interested in the literature and thought of the Renaissance and the Reformation. It was only in the waning days of the Puritan commonwealth that the drive to enforce orthodoxy not only in religion but also in culture, manners, and literature reached the point from which emerged the contemporary stereotype of a Puritan as a zealot and "blue-nose." This latter-day orthodoxy prompted H. L. Mencken to define puritanism as the uncomfortable feeling that someone, somewhere, might be happy.

DEVELOPMENT OF THE COLONIAL SCHOOL SYSTEM

These various elements of early Puritan belief led the New England colonists to establish a school system. So that all members of the congregation could send their children to school, the Puritans developed a system of public schools. These were sectarian, however, since the public supported a single established church and did not allow dissenters' schools to function.[5]

Puritanism was characterized by learned clergy; and, as Merle Curti has pointed out, "A conception of a learned clergy capable of expounding the Bible in the light of scholarship and reason implies a sufficiently well educated laity to follow theological discussions."[6] Not only did the clergy support higher learning and aid in the dissemination of knowledge, Curti goes on to explain, but its most important function was the exposition of God's word. During this period, every denomination relied on revelation as the only certain path to truth and knowledge. In this respect the Bible is all-important, for "God had spoken and His word, contained in the Bible, was holy, absolute, and final." These factors, Curti states, help to explain why Puritan New England led the colonies in compulsory secondary education. As the local school system developed, ministers usually continued to supervise the town schools, and when the towns did not or could not support a Latin grammar school, the ministers prepared ambitious and studious youths for college.

The important role played by clergymen in the development and functioning of the school system in the United States was not, however, confined to Puritans. The Dutch Reformed Church in New York continued its interest in education and schools, and the Germans in Pennsylvania—led by Henry Melchior Muglenberg, the Lutheran leader, and Michael Schlatter, the Swiss-born patriarch of the German Reformed sect—spent a great deal of time establishing church schools. The Presbyterian clergy of the Scotch-Irish also were active in developing educational systems in Pennsylvania, New Jersey, and the Carolinas. Anglican ministers in Maryland and Virginia conducted schools or taught boys in the rectory. The Quakers seemed to be alone among the colonial denominations in placing less emphasis on secondary education. They were somewhat unique because they had no especially trained clergy, since they insisted that any soul, however ignorant and unlearned, could commune directly with the Holy Spirit. Although Quakers tended to deemphasize secondary education, they did not ignore it and actively supported elementary schools.

For various religious, social, and economic reasons, the development of public schools in the North had no exact corollary in the southern colonies. A not untypical statement of southern elitist views was expressed by Virginia's colonial governor William Berkeley in 1671: "I thank God that there are no free schools or printing and I hope we shall not have them these hundred years; for learning has brought disobedience and heresy and sect into the world."[7] It is largely through the efforts of black and carpetbagger legislators governing the southern states in the period immediately following the Civil War that a general system of public education in the South came into being.[8]

This background demonstrates not only that the origins of the U.S. public school system antedate the principle of separation of church and state but also that public education may owe its very existence to the fact that it did.[9] Moreover, this "respectable beginning" may be the reason that the public schools of the United States have been able to maintain a status seldom accorded the state-supported schools of England until recently.

EARLY SCHOOL LAWS

Various statutes and state constitutions confirm that the motivating force behind public education in the American colonies was to enable the students to read the Bible and to become better versed in Protestant religious dogma. The first school law emphasized this aspect of education. In 1642, Massachusetts invested local school boards with the power to "take account from time to time of all parents and masters, and of their children, concerning their calling and implyment of their children, *especially of their ability to read and understand the principle of religion and the capitall lawes of this country.*"[10] (Emphasis added.)

Five years later, Massachusetts law became even more specific. The school law of 1647 stated that every town with more than fifty householders must provide a schoolmaster, so that the children might learn to read the Scriptures. The law quaintly explained:

> It being one chiefe project of ye ould deluder, Satan, to keepe men from the knowledge of ye Scriptures, as in formr times by keeping ym in an unknowne tongue, so in these lattr times by perswading from ye use of tongues, yt so at least ye true sence and meaning of ye originall might be clouded by false glosses of saint seeming deceivers, yt learning may not be buried in ye grave of our fathrs in ye church and commonwealth.[11]

Subsequently, the First Massachusetts Constitution of 1780 took explicit notice of the important role religion must play in public education:

> As the happiness of a people, and the good order and preservation of civil government, essentially depend on piety, religion and morality; and as these cannot be generally defused through a community but by the institution of public worship of God, and of public instructions in piety, their happiness, and to secure the good order and preservation of their government, the people of this commonwealth have a right to invest their legislature with power to authorize and require, and the legislature shall from time to time authorize and require, the several towns, parishes, precincts and other bodies politic, or religious societies, to make suitable provision at their own expense, for the institution of the public worship of God, and for the support and maintenance of public Protestant teachers of piety, religion, and morality, in all cases where such provisions shall not be made voluntarily.
>
> And the people of this commonwealth have also a right to and do, invest their legislature with the authority to enjoin upon all the subjects in attendance upon the instruction of the public teachers aforesaid, at stated times and season, if there be any on whose instructions they can conscientiously attend.[12]

In short, it is clear the Bible played an important part in colonial public schools through reading exercises and for moral instruction. As E. P. Cubberley has noted, "The most prominent characteristic of all early colonial schooling was the predominance of the religious purpose in instruction. One learned chiefly to be able to read the catechism and the Bible. . . . There was scarcely any other purpose in the maintenance of elementary education."[13]

COLONIAL TEXTBOOKS

It should come as no surprise in the light of these antecedents that textbooks in the colonies were heavily influenced by the Bible and religious writings. Indeed, as Merle Curti has pointed out:

> The Bible was to be found in almost every Calvinist household that possessed any books at all, and it was read not only once, but over and over again. The obligation to read it was the chief reason for universal elementary education in communities dominated by Calvinism. Children sometimes learned their first letters from its pages; and even when they got their start in the catechism, a book of piety, or the highly biblical *New England Primer*, they were soon graduated to the Testaments.[14]

Even when the textbook used in elementary schools was not the Bible, it was usually a handbook closely paralleling biblical teaching and drawing heavily upon the Scriptures for its moral lessons. *The Hornbook,* one of the first texts, had at least half its contents taken up with the Lord's Prayer. *The New England Primer* (ca. 1690) mentioned above was noted for its religious flavor in both its rhyming alphabet and its catechism, which were called "Spiritual Milk for American Babes Drawn Out of the Breasts of Both Testaments for Their Soul's Nourishment."[15]

The Bible was the principal reading matter in the German schools that the German Reformed and Lutheran bodies maintained in the Middle Colonies. The children of many of these households could repeat by heart a truly impressive number of scriptural verses. The Bible and Pastorius's *New Primer,* a thoroughly scriptural handbook, were also used as mainstays in instructing youths who were to become Quaker or Mennonite readers. Anglo-Americans had to depend upon imported Bibles until 1777, when the first American edition of the New Testament in English was printed. As early as 1743 the German population could read the Bibles that were being issued from the press of the Pennsylvania Dunker Christopher Sauer.[16]

Early laws in a number of southern states also stressed the important role religion must play in education. A South Carolina statute of 1710 provided "that a free school be erected for the instruction of the youth of this province in Grammar, and other arts and sciences and useful learning and also in the principles of the Christian religion."[17] As late as 1766 North Carolina stated in the preamble of its constitution: "Whereas a number of well-disposed persons, taking into consideration the great necessity of having the proper school of learning established whereby the rising generations may be brought up and instructed in the principles of the Christian religion. . . ."[18]

In spite of the great emphasis placed on Bible reading in home, church, and school by all denominations in colonial America, there were wide differences over scriptural interpretation of the Bible. The vindictive form of these differences was particularly noticeable in the various interpretations concerning the most effective way to achieve salvation. Questions of biblical interpretation were at the heart of most of the sectarian disputes of the time, and each sect used different parts of the Scripture to enforce its own belief. These disputes immediately affected the schools:

The Protestant churches, thus held apart by class as well as by doctrinal differences, believed firmly in the absolute correctness of their own interpretation of scripture. Sectarianism was bitter and this fact had and continued
to have a marked effect on education. It meant that sects regarded education as of first importance in the maintenance of sectarianism; the education of children was to be controlled either in sectarian schools or, as in
New England, in public schools whose policies and practices were determined by the Orthodox in the community.[19]

In reviewing this period of American history, we cannot escape the conclusion that the Bible played an important role in sectarian religion during the
colonial period and that various parts of it were used by the sects to enforce
their beliefs both in and out of the schools. The impetus behind the public
school movement was to maintain the status quo of the established church.
One of the best ways to do this, it was believed, was to have pupils study the
Bible. So pervasive and deep-seated was this view that as late as 1846 the majority of Connecticut public schools, for example, still used the Bible as a
reading book, even though the Connecticut Constitution of 1818 had attempted to divorce the public schools from the influence of the Congregational
Church.

REVOLUTIONARY PRECEDENCE FOR THE NONSECTARIAN PUBLIC SCHOOLS

Amid the sharp sectarian clashes over the public schools in the waning
days of the Puritan theocracy in New England it is possible to discern a slight
movement away from the teaching of sectarian religion. There were a number
of reasons for this change, a principal one being that the Puritan commonwealth and its doctrines, which could flourish only in isolation, no longer
remained isolated when the New England settlements became actively engaged
in commerce and business. The corresponding growth of villages into cities
brought together people of diverse religions seeking employment. This influx
of city workers, coupled with the granting of agricultural land in fee simple to
farmers made the orthodox Puritans a minority group in many areas.

When the majority of people in an area found that their children were
subjected in school to the indoctrination of Puritan religious views (that in
many cases ran contrary to parents' beliefs), the public demanded that such
practices be eliminated. Since a great variety of sects existed in the colonies at
that time, no common denominator could be found that would allow any
religious instruction and still not antagonize some group. As a result, the
logical alternative was to eliminate all sectarian instruction, as finally occurred
in the early days of the nineteenth century. Although vestiges of sectarian instruction remained in the public schools (particularly in New England) well into the nineteenth century, James Lowell in his tribute to the village schools of
Massachusetts foresaw the eventual trend toward secular instruction:

Now this little building, and others like it, were the original kind of
fortification invented by the founders of New England. They were the Mar-
tello towers that protect our coast. *This was the great discovery of our*

> *Puritan forefathers. They were the first law givers who clearly saw and en-*
> *forced practically the simple moral and political truth, that knowledge was*
> *not an alms to be dependent on the chance charity of private men or the*
> *precarious pittance of a trust fund, but a sacred debt which the Com-*
> *monwealth owed to every one of her children.* The opening of the first
> trench against monopoly of church and state; the first role of trammels and
> pothooks which the little Shearfashobs and Elkanahs blotted and blub-
> bered across their copy books, was the Preamble to the Declaration of Inde-
> pendence. . . . What made our revolution a foregone conclusion was the
> act of the General Court passed May 1647, which established the system
> of common schools.[20]

When the Constitutional Convention met in 1787, only five of the thirteen
states still retained some form of an established church. Nonetheless, as Charles
and Mary Beard have argued, secular public education really did not begin to
emerge until the 1830s and was prompted by the effects of Jacksonian
democracy as well as by the influx of substantial numbers of non-Protestant
immigrants.[21]

At the time that the Constitution was ratified and shortly thereafter when
the Bill of Rights was approved, the United States from the standpoint of
religion was a relatively homogeneous nation.[22] Protestantism prevailed.
Nonetheless, serious conflict and acrimonious debate were prevalent enough
among the highly diverse Protestant sects to prompt the framers of the First
Amendment to propose: "Congress shall pass no law respecting an establishment of religion or prohibiting the free exercise thereof." Because contemporary problems are different from these early ones, there is probably some
merit in being cautious about drawing implications for today from the intent
of the Founding Fathers. Even so, the Supreme Court, other policymakers, and
scholars must and do seek to determine the intention of the drafters of the
supreme law of the land.

FRAMING THE FIRST AMENDMENT

Since the Supreme Court's decision in *Everson* v. *Board of Education* in
1947, a furious debate has arisen over what the founding fathers intended
when they framed the First Amendment's religion clauses.[23] If possible, the
debate has intensified since the Court's action in *Engel* v. *Vitale*[24] declaring
state-sponsored nondenominational prayers in the public schools unconstitutional and since the more recent decision in the *Schempp* case in 1963[25] outlawing Bible reading exercises in the public schools as repugnant to the First
Amendment.

In general, the nation falls into two camps concerning what the framers
of the Constitution intended, especially with respect to the establishment
clause. Some maintain that James Madison and the others responsible for the
First Amendment intended to erect an insurmountable wall of separation between church and state. This would prevent any single or multiple establishment from becoming dominant and would prohibit all forms of government
cooperation and indirect aids to religious groups.[26] Others argue that the
draftsmen of the First Amendment did not expect the state to be neutral in

religious matters. The amendment, as they see it, was merely aimed at preventing the establishment of the single state church and not at prohibiting incidental or indirect aid to all religions.[27]

Even a cursory examination of the writings of some of the more seminal thinkers involved with the Constitution and the Bill of Rights at that time, such as John Adams, Thomas Jefferson, and James Madison, reveals a deep suspicion of organized religion, particularly about the effects it might have on education.[28] Jefferson's leading biographers suggest that he believed the First Amendment would create a high and impregnable wall of separation between church and state. James Madison's leading biographer, I. Brant, recognizes that he like Jefferson strongly opposed all religious programs that received even the most indirect public aid: "His aim was to strike down financial aid to religious institutions out of the public purse. . . . Not by the most microscopic concession would he deviate from absolute separation between the authority of human laws and the natural rights of man."[29]

THE DECLINE OF ORTHODOXY

The attacks against sectarian influence in the public schools that began in earnest in the early nineteenth century took two forms. On one hand, many states adopted constitutional and statutory provisions forbidding appropriations from the public treasury for the support of religious instruction in any manner.[30] Others adopted legislation prohibiting the use of sectarian reading and teaching materials in the public schools.

The battle to eliminate sectarian programs from public schools in the United States and to replace them with exercises that were thought to be nonsectarian but might have religious overtones was led by Horace Mann, regarded by many as the father of modern American public schools. Mann believed, for example, that Bible reading programs in the schools, if conducted without comment, would permit children to study the common elements of Christianity and could not possibly injure the religious sensibilities of anyone. The paradox of this position satisfied no one. Indeed, it outraged many of the leaders of the Protestant majority and did little to assuage the irritation of Roman Catholics over Protestant influences in the public schools, which was largely responsible for the creation of Roman Catholic parochial schools.[31]

In the period immediately after the Civil War, when the nation was engulfed by a great wave of immigrants holding divergent religious beliefs, the country's Protestant homogeneity came to an end. Subsequent debates over sectarian instruction and religiously oriented exercise in the public schools grew more heated and took on a national scope, so that by the latter half of the nineteenth century this controversy became a key issue in the national political campaigns.

The Roman Catholic minority in the United States had continued to grow, and their opposition to Bible reading and other exercises that they considered to be Protestant inspired became increasingly hostile. Fear of Catholic domination and public concern over the rumor that the Pope might move the papal see to the United States because of encroachments on papal authority by the

Italian patriot Giuseppe Garibaldi caused a wave of anti-Catholicism to sweep the nation in the early 1870s.[32]

When the eminent cartoonist Thomas Nast was drawn into the debate, he produced a series of biting caricatures attacking papal influence in the United States and criticizing Catholic opposition to the public schools, particularly their stand on Bible reading. It is not surprising that before long the two major political parties became partisans in the dispute. Rightly so or not, the Democratic Party came to be associated with the Roman Catholic cause, which in the public mind stood for state support of parochial schools, opposition to Bible reading in the public schools, and a hostility to public schools in general. As might be assumed, the Republican Party took the opposite view.

This political division on educational issues helps to explain Grant's famous speech to the Army of Tennessee at Des Moines in September 1875, in which he said:

> Let us all labor to add all needful guarantees for the security of free thought, free speech, free press, pure morals, unfettered religious sentiments, and of equal rights and privileges to all men irrespective of nationality, color or religion. Encourage free schools, and resolve that not one dollar appropriated for their support shall be appropriated to the support of any sectarian school. Resolve that neither the state nor the nation, nor both combined, shall support institutions of learning other than those sufficient to afford every child growing up in the land the opportunity of a good common-school education, unmixed with sectarian, pagan, or atheistical dogma. *Leave the matter of religion to the family altar, the church, and the private schools supported entirely by private contributions.* Keep the church and state forever separated.[33]

Moreover, in his annual presidential message to Congress in 1875, Grant recommended a constitutional amendment forbidding the teaching in the public schools of any sectarian religious tenets and further prohibiting "the granting of any school funds or school tax or part thereof, either by the legislative, municipal or other authority, for the benefit or aid, directly or indirectly, of any religious sect or denomination."[34]

Acting on Grant's suggestion, James G. Blaine proposed the following amendment to the United States Constitution in 1876:

> No state shall make any laws respecting an establishment of religion or the free exercise thereof; and no money raised by taxation in any state for the support of public schools, or derived from any public funds thereof, nor any public lands devoted thereto, shall ever be under the control of any religious sect or denomination, nor shall any money so raised or lands so devoted be divided between religious sects and denominations.[35]

The amendment, in a form considerably stronger than that proposed by Blaine — it carried additional provisions prohibiting teaching of religion in publicly supported schools — was overwhelmingly passed by the House (180–7).

However, it failed to obtain a two-thirds majority vote in the Senate (28–16). The vote followed straight party lines. The Republicans voted solidly for the amendment and the Democrats opposed it. One writer believes that the amendment failed because many legislators believed that the state constitutions were adequately equipped to handle this question.[36] Time, however, has suggested that this was not true.

Growing Resistance to Sectarianism in the Public Schools

The contemporary awareness of and resistance to religiosity and sectarianism in the public schools really originated in the late 1870s. At that time, Roman Catholics and Jews were joined by the so-called liberal movement in a vehement attack on all traces of religion in American life. This group attempted to exclude from public places not only the Bible but references to the deity.[37] Shortly thereafter Roman Catholics and Jews gained the support of a group of ministers whose theological beliefs had been shaken by the disproving of many tenets that the authoritarian religion of their youth had insisted must be accepted without question. They concluded that nothing could be gained by forcing religion or any of its manifestations on pupils and teachers in the public schools. A required exercise, they stressed, did not best serve the interests of religion itself. Thus by the end of the nineteenth century an increasing number of people from all sects and with all intellectual leanings had come to regard the practice of Bible reading and related activities as sectarian instruction in the public schools.

As a result, during the late nineteenth and early twentieth century, a number of actions were brought into the state courts seeking to prohibit Bible reading, prayers, and related activities in the schools.[38] It is important to recognize that as the courts heard these cases they failed to agree on whether such exercises or practices in the schools were indeed sectarian. Nonetheless, because of their religious diversity, most Americans by this time had come to believe that public schools should be kept free from sectarian instruction and that all parochial schools should be divorced from public funds. The only way to keep the local, state, and federal governments from becoming embroiled in religious controversies — which had the habit of centering around the public schools — was to keep government as aloof as possible from religion.

This tendency is reflected in specific provisions incorporated into the constitutions of many states that entered the union from around the middle to the end of the nineteenth century, as well as in additions or revisions to the then existing state constitutions and statutes. In 1896 and 1897 the federal government took notice of the prevailing climate of opinion in the nation by announcing in the Appropriation Acts for the District of Columbia for those years a national policy that conformed with the constitutional provisions of the vast majority of the states. The act read in part:

> And it is hereby declared to be the policy of the government of the United
> States to make no appropriations of money or property for the purpose of

founding or maintaining, or adding by payment for services, expenses or otherwise, any church or religious denomination, or any institution or society which is under sectarian or ecclesiastical control.[39]

This avowal of principle, interestingly enough, was not interpreted by the schools to prohibit the reading of the Bible in the District of Columbia public school system, for Bible reading and related practices had long been the rule rather than the exception there prior to the *Schempp* decision.[40]

RELIGION, EDUCATION, AND THE COURTS

Although religious controversies of one sort or another are common in American history, relatively few have received the Supreme Court's scrutiny. Almost all these disputes have been evaluated only in the last decades. Actually, the great bulk of law derived from judicial decisions regarding the First Amendment's provisions on religion has developed since 1940. Startling though it may be, not until the 1960s did some of the crucial issues affecting the most common practices of a religious nature in the public schools come before the Court for review.

It is important to recognize that the *Scopes* case, which is crucial to this study and received such national and international attention, never did receive Supreme Court review. That case was decided by the Tennessee Supreme Court in 1927, but it was not until the *Epperson* case was debated in 1968 that the statute of Arkansas—which was an adaptation of the Tennessee statute (which John Thomas Scopes was convicted of violating in the famous "monkey law" case)—was struck down by the Supreme Court. It is also interesting to note that Scopes's conviction was reversed on the nonfederal ground that the trial judge levied a $100 fine against him that the jury had failed to assess in accordance with state law.[41]

The successor to the statute in *Scopes* was struck down in *Daniels* v. *Waters*.[42] As in the *Epperson* case, the *Daniels* court found impermissible religious purpose. Two years later in 1977, the Indiana Superior Court held that the Indiana Textbook Commission's adoption for use in public schools of a biology text titled *A Search for Order in Complexity*, which described both biblical creation and Darwinian evolution but presented only the former in a favorable light, violated the First Amendment's establishment clause.[43] The court noted the facile neutrality of the text but insisted that the court was bound to determine the purpose of both the book and its inclusion in the curriculum and thought it clear that the purpose was "the promotion and inclusion of fundamentalist Christian doctrines in the public schools."

The Court in *Daniels* noted that the text's "asserted object . . . to present a balanced or neutral argument is a sham that breeches the 'wall of separation' between church and state voiced by Thomas Jefferson." The Court went on to say, "any doubt of the text's fairness is dispelled by the demand for 'correct' Christian answers demanded by the teacher's guide."[44]

What we see reflected here is the classic dilemma between those who insist that their religious freedom requires the handling of certain materials in the public schools and those who insist that to permit such practices with religious overtones is to violate the establishment clause. The Court thus is regularly placed in the position of having to decide which of the two clauses should prevail in instances becoming increasingly more common, in which it is clear on the face of the dispute that the two do indeed involve contradictory rights. Interestingly enough the Court in its earlier decisions was unable to see or did not allude to the problem of the clash of the two religion clauses that regularly confronted it.

Of the two clauses relating to religion found at the outset of the First Amendment, it is probably more meaningful to discuss the judicial evolution of the freedom of religion clause before evaluating the establishment clause, even though the latter appears as the first clause in the Bill of Rights. The rationale behind this approach is that the Supreme Court on balance has had less difficulty with defining the freedom of religion clause than it has had with the First Amendment clause dealing with the establishment of religion.

Concerning the freedom of religion, the Court has devised a frame of reference that views such freedom on three different levels — the right to believe, the right to advocate religious beliefs, and the right to practice one's religious beliefs. The degree of protection a person receives will vary on each level, the Court has made clear.

The right to believe, for example, is absolute and cannot be abridged. To date, at least, every person has always been held to be secure in the privacy of his own thoughts and beliefs. Although relatively unrestricted, the right to advocate one's religious beliefs can be curtailed, according to the Court, if there is immediate danger of substantial injury to others. It is not enough that such advocacy result in mere injury to others; it must result in substantial injury. For example, the Court might accept a regulation prohibiting religious advocacy that appears to incite the violent elimination of members of other religious sects.

The right to practice one's religion is viewed differently. Less freedom is authorized by the Court for the *practice* of religion than for the right to advocate it, and this is certainly much less freedom than is permitted for the right to *believe*. This distinction was made clear in the so-called *Mormon* cases of the 1880s, which were some of the earliest freedom of religion cases to come before the Court. In *Reynolds* v. *U.S.* certain Mormons challenged the constitutionality of the federal statute governing the territory of Utah that made bigamy a crime.[45] They alleged that this law violated the freedom of religion guaranteed by the First Amendment, since a basic tenet of the Mormon faith was a belief in plural marriages.

The Court ruled that the free exercise of religion guaranteed in the First Amendment is an absolute right only with respect to beliefs. The Congress could, however, as a valid exercise of its duty to protect the health, welfare, and morals of the nation, impose reasonable regulations that might have the effect of restraining certain religious practices. In essence, the Court said that while the Mormons could *believe* that they could or should have plural marriages, they simply could not *practice* these beliefs in this country. More recently, the same judicial reasoning was used to reject the charge by certain southern religious sects who handle poisonous snakes during religious services that state laws outlawing these practices violate the freedom of religion.

In the 1960s the Court rejected the argument that so-called blue laws (laws that made it a crime to conduct business on a Sunday) were a violation of the freedom of religion guaranteed by the First Amendment.[46] In 1961 the Court held that although these laws originally might have been enacted to force people to attend church, the contemporary purpose behind them was secular rather than religious since the intent was to designate one day for rest, relaxation, and family companionship. In the same year, however, the Supreme Court struck down the state of Maryland constitutional provision that required an oath in the belief in the existence of God as a prerequisite for holding public office.[47]

Two years later in 1963, the Court was confronted with another classic conflict between the two religion clauses when it reviewed the *Sherbert* v. *Verner* case.[48] There the State of South Carolina had ruled that a Seventh-Day Adventist who refused to work on Saturday (her sabbath day) and was discharged was not entitled to unemployment compensation benefit payments. The Court rejected the argument that for the state to provide such benefit payments would constitute establishing a religion in violation of the first clause of the First Amendment, since it would single out for preferential treatment the Seventh-Day Adventist religion. The Court held instead that the denial of benefits to Sherbert constituted infringement of her constitutional right of the freedom of religion because her disqualification imposed a burden on her free exercise of religion by forcing her to choose between adhering to her religious precepts and forfeiting benefits for abandoning them.

In 1981 the Court reaffirmed the *Sherbert* decision in the case of *Thomas* v. *Review Board of Indiana Employment Securities*.[49] There the Court held that a Jehovah's Witness was eligible for employment benefits even though he quit his job because he was shifted to a position that required him to deal with the manufacture of war materials.

As for the dispute over creationism, one of the most important recent cases reviewed by the Supreme Court is that of *Wisconsin* v. *Yoder* in 1972.[50] The question before the Court was whether the Amish, given their well-established belief that education beyond the eighth grade teaches worldly values at odds with their religious creed, could be compelled to attend school without violating their rights guaranteed by the freedom of religion clause. The Court for the first time held that a religious group could be immune from compulsory attendance requirements in the public schools. Chief Justice Warren Burger found a violation of Amish free exercise of religion in the Wisconsin compulsory school law, but he made it clear that this holding would not apply to "faddish new sects or communes."

At present a number of fundamentalist sects are attacking various state compulsory school laws on the basis of this decision. These religious groups have established their own schools, which do not meet many of the basic requirements of state school laws. The cases have not as yet come to the Supreme Court for a full-fledged review.

As these recent cases demonstrate, the practical difficulty that the Court faces is that it must choose between one or the other of the contending clauses dealing with religion in the Constitution. The precise nature of the dilemma was conveyed by Justice Potter Stewart in his concurring opinion in *Sherbert* v. *Verner:*

> The case presents a double-barreled dilemma which in all candor I think the Court's opinion has not succeeded in papering over. . . .
>   To require South Carolina to so administer its laws as to pay public money to the appellant under the circumstances of this case . . . [is] clearly to require the state to violate the establishment clause as construed by this court.[51]

In their dissents in the same case, Justices John Marshall Harlan and Byron R. White also questioned upholding the right of the Seventh-Day Adventists to receive unemployment benefits because of their religious inability to work on Saturday: "The meaning of today's holding . . . is that the state *must single out* for financial assistance those whose behavior is religiously motivated, even though it denies such assistance to others whose identical behavior (in this case inability to work on Saturdays) is not religiously motivated."

THE ESTABLISHMENT OF RELIGION

In 1962, Justice Hugo L. Black, in speaking for the Court in the *Engel* v. *Vitale* case prohibiting prayer programs in the public schools, addressed himself to the distinction between the establishment clause and the free exercise clause. He noted that although the two clauses might overlap in certain instances, they nonetheless forbid two quite different encroachments upon religious freedom: "The Establishment Clause, unlike the Free Exercise Clause does not depend upon any showing of direct government compulsion and is violated by the enactment of laws which establish an official religion whether those laws operate directly to coerce non-observing individuals or not."[52]

Then in 1963 in the landmark *Abington School District* v. *Schempp* case striking down Bible reading exercises in the public schools, the Court fashioned a test to determine whether a state law or practice violated the establishment clause. The test — as Justice Tom C. Clark, speaking for the Court, saw it — is: "What are the purposes and primary effect of the enactment?" The First Amendment is violated, the Court announced, "if either the purpose or primary effect of the law is the advancement or inhibition of religion." To clarify this position, the Court emphasized again the distinction between the establishment clause and the free exercise clause, and noted that a given action might violate one but not the other: "A violation of the Free Exercise Clause is predicated upon coercion while the Establishment Clause violation may not be so attended."[53]

In a series of recent cases, the Supreme Court has been consistent in insisting that a state-supported, religiously oriented practice must meet three requirements in order to square with the demands of the establishment clause of the First Amendment. First, it must have a secular purpose; second, it must have a primary secular effect; and third, it must not have excessive religious entanglement.[54]

In *Stone* v. *Graham,* decided in 1980, however, the Court appears to be revising these rules concerning the establishment clause, or at least may have initiated substantial internal argument over their future applicability. Although the majority clung to the established judicial criteria concerning the prohibition against religiously oriented practices, four justices dissented. Of the four dissenters, at least two—Justices William H. Rehnquist and Potter Stewart—urged overturning the fundamental legal principles controlling judicial resolution of issues involving the establishment clause. The other two dissenters, Chief Justice Burger and Justice Harry A. Blackmun, while noting they were dissenting because they would have given the case plenary consideration rather than the summary treatment afforded by the majority *per curiam* opinion, may have done so because they wished to see reopened a debate on the fundamental issues of prevailing principles of law. Otherwise, it might be argued, they should have been content to accept the majority opinion following the doctrine of *stare decisis* supportive of existing principles affecting church, state, and education.

In the *Stone* v. *Graham* case, the Supreme Court struck down a Kentucky statute requiring that a copy of the Ten Commandments be posted in every public school. The copies, which the statute decreed were to be sixteen inches wide by twenty inches high, were to be purchased with private contributions. The statute also required that "In small print below the last Commandment shall appear a notation concerning the purpose of the display as follows: 'The secular application of the Ten Commandments is clearly seen in its adoption as the fundamental legal code of Western Civilization and the Common Law of the United States.' " The majority opinion reiterated the three-part test that had been developed over the years to determine whether a challenged statute is permissible under the establishment clause of the First Amendment: "First the statute must have a secular legislative purpose; second, its principal or primary effect must be one that neither advances nor inhibits religion . . . ; finally the statute must not foster an excessive government entanglement with religion."

The Court went on to explain that if a statute violated *any* of these three principles, it must be struck down as violative of the establishment clause. The Court concluded that the Kentucky statute had no secular legislative purpose, and therefore it was unconstitutional. The Court recognized that the state of Kentucky insisted that the statute served a secular legislative purpose in that the legislature required the small-print notation described above. The Court also observed that the trial court in Kentucky found the avowed "purpose" of the statute to be secular even though the Kentucky court labeled the statutory declaration "self-serving."

The Supreme Court, however, found that the so-called avowed secular purpose was not sufficient to avoid conflict with the First Amendment. It noted that in the *Schempp* case of 1963 the Court had held unconstitutional the daily reading of Bible verses and the Lord's Prayer in public schools despite the school district's assertion of such secular purpose as "the promotion of moral values, the contradiction of the materialistic trend of our times, the perpetuation of our institutions and the teaching of literature."

In Kentucky's case, the Court found that the preeminent purpose for posting the Ten Commandments on the schoolroom wall was "plainly religious in nature." The Court explained that the Ten Commandments are undeniably a sacred text in the Jewish and Christian faith and that no legislative recitation of a supposed secular purpose can blind the Court to that fact. Moreover, the Court explained that the commandments did not confine themselves to arguably secular matters such as honoring one's parents, killing or murder, adultery, stealing, false witness, and covetousness. Rather, the first part of the commandments concerned the religious duties of the believers: worshipping Lord God alone, avoiding idolatry, not using the Lord's name in vain, and observing the Sabbath.

Furthermore, the majority noted that in the Kentucky situation, the Court did not find a case in which the Ten Commandments were integrated into the school curriculum, where the Bible might constitutionally be used in a study of history, civilization, ethics, comparative religion, or the like, as the Court had noted were perfectly acceptable practices in the *Schempp* case. Posting a religious text on the wall, the Court observed, certainly serves no such educational function. If the posted copies of the Ten Commandments were to have any effect at all, the Court noted, it would be to induce the schoolchildren to read and meditate upon and perhaps to venerate and obey the Ten Commandments. Although this might be desirable as a matter of private devotion, the Court said such practices are not a permissible state objective under the establishment clause.

In the *Stone* case, which was decided by a 5–4 vote, and in the 1981 case of *Thomas* v. *Indiana Review Board,* Justice Rehnquist, who some see as an emerging leader of the so-called right-wing bloc of the Supreme Court, sharply criticized the majority's views and went as far as to urge the total abandonment of judicial doctrines of church-state separation going back to their beginnings in the *Everson*[55] case in 1947. As he explained in the *Thomas* case:

> It might be argued that cases such as *McCollum* v. *Board of Education; Engel* v. *Vitale; School Board* v. *Schempp; Lemon* v. *Kurtzman;* and *Committee for Public Education* v. *Nyquist* were all wrongfully decided. [The] aid rendered to religion in these latter cases may not be significantly different in kind or degree than the aid afforded Mrs. Sherbert or Thomas. For example, if the state in *Sherbert* could not deny compensation to one refusing to work for religious reasons, it might be argued that a state may not deny reimbursement to students who choose for religious reasons to attend parochial schools. The argument would be that although a state need not allocate funds to education, once it had done so it may not require any person to sacrifice his religious beliefs in order to obtain an equal education.[56]

A recent example of the legal skirmishing over attempts to require the teaching of creationism in the public schools occurred in Arkansas, where a federal district court on January 5, 1982, struck down a recently enacted state statute entitled the Balanced Treatment for Creation-Science and Evolution-Science Act.[57] The essential mandate of the law as the court saw it was to ensure that "Public schools within this state shall give balanced treatment to creation-science and to evolution-science."

The statute, which was criticized for creating an "establishment of religion" in violation of the First Amendment, was attacked by a dazzling array of prominent church and educational groups. These included the resident Arkansas bishops of the United Methodist, Episcopal, Roman Catholic, and African Methodist churches, the principal officials of the Presbyterian churches, and other Southern Baptist and Presbyterian clergy. Organizational plaintiffs who joined the attack included the American Jewish Congress, the Union of American Hebrew Congregations, the American Jewish Committee, the Arkansas Education Association, and the National Association of Biology Teachers.

At the outset, the court recognized that the role of religious fundamentalism was crucial to the drafting and the enactment of the statute under review. The court found that a central premise of fundamentalism "has always been a literal interpretation of the Bible and a belief in the inerrancy of the Scriptures." In addition, one aspect of fundamentalist political efforts, particularly in the South, was the promotion of statutes prohibiting the teaching of evolution.

The court also found that "Creationists view evolution as a source of society's ills, and the writings of [Henry M.] Morris and [Martin E.] Clark are typical expressions of that view." To emphasize this point, the court noted that on page 31 of *The Bible Has the Answer* Morris and Clark conclude that "evolution is thus not only anti-Biblical and anti-Christian, but it is utterly unscientific and impossible as well. But it has served effectively as the pseudo-scientific basis of atheism, agnosticism, socialism, fascism and numerous other false and dangerous philosophies over the past century."

Upon reviewing the legislative history of the statute in question, the court found that it failed the first requirement of the three-pronged test formulated by the U.S. Supreme Court to determine whether a challenged statute violated the establishment of religion clause. The Arkansas law lacked a "secular legislative purpose" since the court found background details that the "author of the Act had publicly proclaimed the sectarian purpose of the proposal" and the state residents "who sought legislative sponsorship of the bill did so for a purely sectarian purpose." Furthermore, as the court saw it, the bill's sponsors made no effort to consult or discuss the matter with any educators or scientists before launching what the court refers to as their "unprecedented intrusion in school curriculum."

At the time of this writing no appeal has been taken from the federal district court's decision. But various representatives for fundamentalist groups have indicated that they expect more supportive action in several other states in which similar measures are under legal attack. It seems safe to predict that this area of controversy will see a great deal more litigation in the 1980s.

For Further Reading

Boles, Donald E. *The Bible, Religion, and the Public Schools.* Ames: Iowa State University Press, 1965.
Studies the evolution of the controversy over Bible reading and other religiously oriented pro-

grams in the public schools along with pertinent court decisions and conflicting views of religious and educational groups.

MICHAELSEN, ROBERT. *Piety in the Public Schools.* London: Macmillan, 1970.
Examines the high points in the history of the relationship between religion and the U.S. public schools, focusing on the dynamics of American religion and the possible role of the public school as America's "established church."

MORGAN, RICHARD E. *The Politics of Religious Conflict.* New York: Pegasus, 1968.
Utilizes conceptual equipment of modern political science to analyze the civic conflict called the politics of church and state and to suggest possible developments in the 1980s.

PFEFFER, LEO. *Church, State and Freedom.* Boston: Beacon Press, 1953.
Demonstrates that the origin of the public school not only antedates the principle of separation between church and state but public education may owe its very existence to the fact that it antedated separation.

SCOPES, JOHN T., AND PRESLEY, JAMES. *The Center of the Storm: Memoirs of John T. Scopes.* New York: Holt, Rinehart, 1967.
Combines Scopes's remarkable memory with courtroom documentation to capture the tension and color of the momentous "monkey trial."

STOKES, ANSON P. *Church and State in the United States.* 3 vols. New York: Harper & Row, 1950.
The most ambitious historical study tracing church-state relations in the United States from colonial times to midtwentieth century and demonstrating that the primary arena in which the battles have been fought are the public schools.

[14]

The Battle in Iowa: Qualified Success

JACK A. GERLOVICH AND
STANLEY L. WEINBERG

IOWA is a stable, prosperous, heartland state with an economy based on agriculture and industry. Its seven moderate-sized metropolitan areas are not unduly afflicted by slums, crime, or racism. The state is known for its literacy rate, which is the highest in the nation, and for its temperate politics. Both major parties are strong, and Iowa politics has lively liberal and conservative components.

Although it lacks a large black population, the state otherwise is ethnically diverse. German, Irish, Scandinavian, Czech, and Dutch communities proudly maintain their cultural identities. Iowa has proportionately the largest Vietnamese population of any state. On the initiative of their popular governor, Robert D. Ray, Iowa citizens supported a volunteer medical team in the refugee camps in Thailand. Refugees were brought to Iowa, and their assimilation and adjustment were supported by a state office. The many small towns in which the refugees settled have welcomed them warmly.

In addition to the major liberal, main-line, and evangelical Protestant churches and the Roman Catholic church, Iowa has many smaller religious denominations. These include the pietistic Amana colonies, the Amish, Mennonites, Quakers, Jews, Mormons (both the Salt Lake City and the Reorganized Latter Day Saints churches), and the Maharishi Transcendental Meditation sect. These varied groups are accepted with great tolerance. Indeed, when the Amish protested application of the state's compulsory attendance law to their children, modification of the law to accommodate their demands won general approval.

With its strong traditions of diversity, urbanity, and tolerance, Iowa would hardly seem to offer fertile ground for the growth of an extremist movement or for an ideological conflict. Yet the state appears to have been a major target of the current creationist drive in the United States. In the six-year period 1977–1982 nine creationist equal-time bills were introduced in the state legislature, but none was passed (see Table 14.1). Events in Iowa have been publicized in the national press and have drawn nationwide attention.[1]

189

TABLE 14.1.   SUMMARY OF CREATIONIST BILLS IN THE IOWA GENERAL ASSEMBLY, 1977–1982

Designation of bill	Year	Principal features	Disposition
H.F. 154	1977	Mandated equal time	Died in committee
S.F. 261	1977	Mandated equal time	Died in committee
S.F. 458	1979	Mandated equal time; applied to colleges; also applied to nonstandard theories of medical treatment	Held over to next legislative session
S.F. 458	1980	S.F. 458 made permissive; bill applied to all "public school corporations"	Lost on senate floor, 22–23
S.F. 280	1981	Mandatory; required expenditure for textbooks	Died in committee
S.F. 97	1981	Identical to S.F. 458	No action
S.F. 280	1982	Mandated balanced treatment of creationism and evolution	Died in committee
S.F. 677	1982	Mandated balanced treatment of creationism and evolution; prohibited religious instruction or use of religious materials	Died in committee
S.F. 61	1982	Mandated balanced treatment of creationism and evolution	Died in committee

Thus, throughout the nation, Iowa has come to exemplify proevolution response to creationist initiatives.[2] Educational authorities in half a dozen other states have sought guidance from the Iowa Department of Public Instruction (DPI) in dealing with their own creationist problems. Individuals in other states have similarly been in touch with individuals in Iowa. In many cases Iowa's strategic and tactical examples have been followed.

It is hard to determine why the creationists chose Iowa as a key target—if indeed they did choose the state, and the visibility of creationism here is not just fortuitous. One factor may be the intensive activities of the state's most outspoken creationist, who for years has been importuning the governor, the legislature, the Board of Regents, the DPI, the Iowa Academy of Science (IAS), area education agencies, and other official and semiofficial bodies. Another factor may be the unique situation at Iowa State University (ISU), a highly rated university with outstanding colleges of agriculture, home economics, engineering, and veterinary medicine. The dean of engineering, David R. Boylan, is a leading creationist, an activist in procreationist groups on the ISU campus, and a member of the Advisory Board of the Institute for Creation Research in San Diego. Dean Boylan's activities have been supported by a 400-member Bible Study Association and its offshoot, Students for Origins Research. Some of the ardent creationist students who make up these associations have been involved in repeated creationist-evolutionist confrontations that have drawn the attention of the press, the state legislature, the district court, and the public.[3]

The purpose of this chapter is to show, on the basis of Iowa examples, the types of activities that the scientific, educational, legislative, and lay publics may expect from creationists as well as some mechanisms for addressing such tactics.

INITIAL CREATIONIST EFFORT

In February 1977, House File 154 was introduced into the lower house of the state legislature.[4] It proposed: "If a public school district offers courses which teach pupils about the origin of humankind and which include scientific theories relating to the origin, instruction shall include consideration of the creation theory as supported by modern science." The bill attracted little attention, and it died in committee.

Creationist efforts, however, had just begun in the state. Local newspapers began receiving large numbers of letters to the editor concerning the teaching of creationism in the public schools. Legislators and candidates were persistently lobbied and asked to pledge support for new creationist legislation. During the next three years, public meetings throughout the state promoted the cause of creationism in the schools. Responses from proevolutionists were initially sporadic, uncoordinated, and feeble. Yet it began to be apparent, even to evolutionists, that the creationists were becoming a well-organized, well-briefed, and effective lobby.

DPI Response: The Study

In May 1977, upon the insistence of an ardent local creationist, a local school district made a written appeal to the DPI to "consult with experts in the science community to determine if the evidence used to support the creationist theory was credible and should be made available to students as an example of good scientific investigation." Also in the spring of 1977, a member of the Iowa legislature asked the DPI to study the status of the teaching of creationism in the public schools of other states. State Superintendent Robert D. Benton promptly asked one of us (Gerlovich), the DPI science consultant, to study the issue raised by these requests.

To assure objectivity in drafting guidelines for addressing the controversy, the science consultant believed that input from a diversity of individuals, organizations, and other sources was desirable. The literature was searched.[5] The consultant circulated a questionnaire to his counterparts in the departments of education of the forty-nine other states, inquiring into the status of the teaching of creationism and evolution and into policies for addressing creation-evolution controversies. Forty-five states responded.

It appeared that in most states creationism was not yet a major educational or political issue. Few states therefore had *any* guidelines for addressing the issue. Practice generally involved either neutrality or screening and selection of educational materials by a state committee. Six states required, either by legislation or by departmental regulation, some form of recognition of creationism. Those states reporting an explicit posture on the issue generally made use of position papers developed by their state teacher associations or by national science teacher organizations.

Various national and state science and science education organizations were asked for their policy statements relevant to the teaching of creationism and evolution in public schools. The organizations approached uniformly took the position that creationism was a sectarian belief, and that introducing it into the public schools violated the U.S. constitutional doctrine of separation of church and state. Evolution was supported as meeting the criteria of science.

In November 1977 a survey was conducted of science teachers in attendance at the annual meeting of the Iowa science teachers section of the IAS. Returns showed that most science teachers (90 percent) thought that either a strict separation of church and state should be maintained (42 percent) or guidelines for addressing the controversy should be developed and provided to local schools (48 percent). The schools could then make their own decisions regarding the teaching of creation and evolution. Approximately 10 percent of the responding teachers thought that equal time should be given in science classes to the teaching of creationism and of evolution.

A study of legal precedents showed that court findings generally were premised on the First and Fourteenth amendments to the U.S. Constitution.[6] The courts used three tests to determine the legal standing of statutes at issue under these amendments: the statute must have a secular legislative purpose; the principal or primary effect must be neither to advance nor to inhibit religion; the statute must not foster an excessive government entanglement with religion.

Leaders of Iowa's principal main-line and conservative religious denominations were interviewed. Most of the clergy and theologians stated that religion deals with the "who" and the "why" of ultimate origins, whereas science addresses the "how" of natural phenomena, including origins and biological development. Most of these religious authorities indicated that because of this basic difference in the nature of scientific and of theological concepts (including creationism), the specifics of each discipline should be confined to its respective sphere of activity — science to the school, religion to the home and church.

Iowa science educators, science supervisors, Academy of Science members, religion and science department professors at Iowa colleges, Civil Liberties Union members, and parent-teacher organization administrators were also interviewed. The consensus among these individuals was that creationism was based ultimately on religious belief and was therefore inappropriate for the science curriculum of Iowa's public schools.

To add a third-party perspective to the investigation, university professors of philosophy were approached. They responded that the introduction of creationism into the curriculum would open the door to other pressure groups, lobbyists, and politicians. Their central concern was that limits must be placed on the demands that should be made of public schools.

DPI RESPONSE: THE DECISION

In view of developments in other states, it seemed to DPI administrators that responding to special-interest pressure groups by modifying curricular content through legislative mandate was not an acceptable educational practice. It was thought that the preparation of a DPI position statement incorporating the above findings might better serve all concerned parties. This approach would also encourage local school districts to exercise autonomous control over curricular issues. Such procedure is consistent with long-standing educational policy in the state.

Consequently, a position paper was prepared that passed through seven revisions.[7] It incorporated input from all the groups mentioned above. The

TABLE 14.2. RESPONSES OF KNOWLEDGEABLE READERS TO CONCEPTUAL BASIS AND INTENT OF IOWA DPI POSITION PAPER

Group	Supported DPI position paper	Did not support DPI position paper	Undecided	Total
Iowa scientists, science educators, college and university administrators, science supervisors, textbook authors	68	1	1	70
Iowa religious denominations (conservative and liberal), university and college religion departments	17	8	5	30
State and national science and science education organizations	12	1	3	16
Professors of philosophy at state and private colleges	2	0	0	2
Total	99	10	9	118

final draft was sent to 235 reviewers from the above groups for their reaction and support. Of the 118 respondents, 99 supported the concept and intent of the position paper, 10 could not support it, and 9 were undecided (Table 14.2).

The paper was reprinted in numerous science education journals.[8] Copies were sent to each state department of education throughout the United States and to each school superintendent throughout Iowa. The response was very positive. Between 1978 and 1981 approximately 1,000 additional copies of the paper were requested from several foreign countries, and from diverse science and science education organizations, religious groups, legislators, educational administrators, and laypersons throughout the United States.

The DPI decision on the creation-evolution controversy presented in the operative section of the position paper may be summarized as follows. The public schools cannot be all things to all students. It is the function of science teachers to teach valid science, not religious doctrine or other nonscientific concepts. Evolution is recognized as a well-supported scientific theory. The validity of personal religious beliefs, including belief in creationism, is also recognized; but for counseling and further explanations in these areas, student inquiries should be directed to theological experts or to other appropriate individuals or institutions.

The DPI position is based on the tradition of religious freedom in the United States and on the establishment clause in the First Amendment to the Constitution. In the words of an Indiana court decision: "The prospect of biology teachers and students alike forced to answer and respond to continued demand for 'correct' Fundamentalist Christian doctrine has no place in public schools."[9] The DPI position paper is presented in full in Appendix A to this chapter.

THE LEGISLATIVE CAMPAIGN, 1978–1979

During 1978 and 1979, meetings dealing with the controversy were held at universities, colleges, the Capitol building and elsewhere in the state. The *Des Moines Register,* which has a statewide circulation, carried a large and continuing volume of correspondence. The *Register* reported that it was receiving a disproportionate number of letters supporting creationism versus those supporting the teaching of evolution. But the newspaper printed approximately equal numbers of letters on both sides. The letters covered almost all aspects of the issue. Editorially the *Register* supported evolution and opposed equal time.[10]

In February 1979 a new bill, Senate File 261, was introduced into the Iowa Senate,[11] proposing that "Whenever the origin of man or the origin of the earth is alluded to or taught in the educational program of public schools of this state, the concept of creation as supported by scientific evidence shall be taught as one theory." The bill stimulated renewed discussion throughout Iowa, among both professional and lay persons. Governor Ray said he was against a state mandate that public schools in Iowa must also teach creation theory if evolution theory is taught. In a news conference Ray stated that school officials already had the flexibility to address the subject of creation and should continue to have this control.[12]

Still another bill, Senate File 458, was introduced in March. This bill stated: "Wherever the origin of humankind, or the origin of the earth is taught in the educational program of public schools of this state, the concept of creation as supported by scientific evidence shall be included."[13] The bill had nine amendments attached to it, including two that created a great deal of concern at the college level. The first read: "Insert after the word 'schools' the words 'community colleges, merged area schools, or universities.' " The second alarming amendment read:

> Whenever the theory of the medical treatment of diseases of humankind is taught in the educational program of the public schools, community colleges, merged area schools, or universities of this state, the concept of treatment of the diseases of humankind by osteopathic, chiropractic, Christian Science, and faith healing, as supported by scientific evidence, shall be included.

Projected Cost of an Equal-Time Program

In February 1979 the legislature asked the DPI to prepare a "fiscal note" estimating the cost of introducing creationism into all public school science classes. The outline of Fiscal Note 79-269 that follows includes its essential components: basic operational assumptions; derivation of estimates; operating expenses.[14]

Assumptions

1. That no new state funds would be allocated to implement this law.

2. That creationism would be taught as one alternative theory whenever the origins of the earth or man were discussed; therefore, materials would be required by every teacher of science from kindergarten through the twelfth grade.

3. That in some form, evolutionary theory concerning the origins of, and changes in, the earth and man is included in all science textbooks.

4. That currently no teachers of science in the public schools are adequately prepared to teach creationism.

5. That Iowa public schools do not now possess instructional materials for teaching creationism.

6. That in-service programs at the Area Education Agency level would be required to train present teachers of science from kindergarten through grade 12.

7. That teacher in-service would occur during the contract year.

8. That staff members of regents institutions would need to be prepared to train future science teachers in the teaching of creationism.

Derivation of Estimates for Personal Services

Travel to in-service courses for 18,200 teachers, 4 teachers/car, 100 miles round trip average, local education agency to area education agency at $0.15/mile	$ 68,250
Owing to large numbers of teachers involved the average number of area education agency in-services would be two	68,250
Substitute teacher cost for 18,200 teachers of science in kindergarten through twelfth grade at $30/day	546,000
Consultant fee to conduct area education agency in-service workshops, $150/day (two sessions/area education agency)	4,500
Travel cost, $500 × 15 area education agencies	7,500
Subtotal	$ 694,500

Operating Expenses

Textbooks, $10 each × 569,000 students	$5,690,000
Supporting materials, three sets for each of fifteen area education agencies (films, filmstrips, monographs, etc.), $300 each set	13,500
Preparation of regents institutions science methods staff to preservice new science teachers (one at each university), $1,000 each for consultant and materials library	3,000
Subtotal	$5,706,500
Total	$6,401,000

THE SCIENTISTS JOIN THE BATTLE

Iowa scientists had long been nervously watching the expanding creation-evolution controversy without, as a group, taking any visibly active part in it. In April 1979 they were at last galvanized into acting coherently in defense of evolution. Voting members of the board of directors of the Iowa Academy of Science approved the following statement:

> As scientists we object to Senate bill #458 which proposes to equate "scientific creationism" and evolution as scientific theories. We object primarily because "creationism" is not science but "religious" metaphor clothed as "scientific" fact. There is an overwhelming acceptance by knowledgeable scientists of all disciplines that evolution is consistent with the weight of demonstrable evidence.
>
> We feel that Iowa students deserve an education consistent with views of legitimate scientists and that "creationist" views have no proper place in the science classroom. We fully respect the religious views of all persons but we object to attempts to require any religious teachings as science.[15]

The academy statement (see Appendix B) was distributed to members of the senate on the day of a public hearing on Senate File 458 before the Senate Education Committee. It also appeared in the press and apparently had substantial effect. The committee hearing was well attended by both senators and the public. Media coverage was extensive. Students in favor of creationism from ISU held orderly demonstrations. After the hearing, the equal-time bill was referred to the finance committee, where it was held without being reported to the floor.

Several factors probably contributed to the bill's failure to progress, including the substantial discussions of the controversy in the newspapers and elsewhere, the *Register's* editorial position, Governor Ray's stand, the DPI position paper, and effective opposition by proevolution senators. Another consideration was the required expense, pointedly brought to the attention of the cost-conscious legislature by DPI Fiscal Note 79-269. Especially significant was the intercession of the Academy of Science and of a large number of evolutionary scientists acting independently. The scientists were active both in generating proevolution publicity and in speaking at the senate hearing and at other meetings. The active concern that Senate File 458 stimulated among the scientists of Iowa has not since been lost. Instead it has grown and deepened.

THE LEGISLATIVE CAMPAIGN, 1979–1982

During the summer of 1979 an interim study committee of the Iowa legislature was charged with reviewing the controversy. The committee decided not to recommend a creationist equal-time bill to the 1980 legislature. During the same summer one of the sponsors of the 1979 senate bill queried the state attorney general's office concerning purported discrimination against creationism in the schools. The opinion of the assistant attorney general was that "nothing in Iowa law requires the teaching of the creationist model in public school science courses."[16]

Despite the unfavorable report of the study committee, a creationist initiative in January 1980 resulted in a new version of Senate File 458, which proposed: "Whenever the origin of humankind or the origin of the earth is alluded to or taught in the educational program of the public school corporations of this state the concept of creation as supported by scientific evidence may be included."[17] Two interesting changes appeared in this bill: the intent of the legislature was shifted from mandatory to permissive, and introduction of the term "school corporations" could be interpreted to include all levels of education. Following long debate, the bill failed. The vote was 23–22; twenty-six votes were needed for passage. The sponsor switched his vote so that he could file a motion to reconsider at a later date. At the time of this writing the bill has not been called up for reconsideration.

Of the two creationist bills that were filed in the senate in 1981, neither reached the floor for debate and vote. Proponents of Senate File 280 acknowledged in the Education Committee that the bill would require the schools to buy new textbooks.[18] The legislators felt that burdening the already financially troubled schools with new and additional expenses would hardly be a popular move. Further, since the schools already had the right to teach creationism if they wished, the legislation seemed redundant. Senate File 97, sponsored by Senator Raymond Taylor, was identical to the version of Senate File 458 that Senator Taylor had also sponsored and had been defeated a year earlier.

In the 1982 legislative session, three creationist bills, Senate File 677, 280, and 61, were referred to the Senate Education Committee. Two of the bills were reentries of earlier measures, while Senate File 677 was an adaptation of the Ellwanger bill that became law in Arkansas. There was no disposition on the part even of creation-oriented senators to press for hearings or reports on any of the bills. As one senator who had sponsored an earlier equal-time bill said: "No matter what position I take on any of these bills, I make enemies. I hope they all stay in committee." All three bills died in committee. The creationists' legislative drive that began in 1977 seemed to have run out of steam.

Science Supervisors as Targets

The Iowa Council of Science Supervisors (CS²) is concerned with all levels of science education and it advises the DPI concerning science education issues and concerns. Members are familiar with creationist efforts throughout Iowa and the United States. In March 1980 an outspoken creationist in the state asked the president of the CS² for time at the November meeting to present creationist materials. Since members of the CS² were already quite familiar with this material, the request was rejected. The creationist applicant thereupon appealed to sympathetic legislators and officials. Political pressure was exerted on the CS² either to allow this and other such presentations or to justify its refusal to hear them. The CS² stood by its position and the creationist was not heard.

A year later a "Position Statement on the Creation/Evolution Controversy from the Iowa Council of Science Supervisors (CS²)" was approved by 91 percent of the council's voting members.[19] The paper discusses the responsibilities of science educators who are dependent on the scientific community for science content. It further states that teachers should be discouraged from using materials identified by competent scientists as nonscience, prescience, or pseudoscience, except as examples of these categories. The paper appears in entirety in Appendix C.

ROLE OF THE ACADEMY OF SCIENCE

Up to early 1983 no creationist bill had been passed by the Iowa legislature. Scientists affiliated with IAS played an active role in blocking several of the proposed bills. The actions of the academy, however, were often hasty and ad hoc. During 1980 and 1981, Iowa scientists discussed at length how to develop a more consistent and better coordinated defense of evolution in the face of prospective further creationist initiatives.

In February 1981 a statement of the position of the Iowa Academy of Science (Appendix B) was sent to 1,250 members for their approval; of the 725 responses, 92 percent approved the statement.[20] The paper communicated the strong opposition of the IAS to introducing "scientific creationism" into science classrooms. The IAS recognized creationism to be a religious doctrine offered as science. Supernatural explanations for natural events or their origins were held to be contrary to the nature of science and outside its realm.

The position paper is being distributed to Iowa schools that request help in addressing the question of the status of creationism. The paper will also be presented to the state legislature should creationism again become a live issue there.

CREATIONISM AT LOCAL LEVELS

Despite successes in the legislature, substantial victories cannot be claimed for the evolutionary position. The basic strength of the creationist opponents of science does not lie in winning legislative victories, or even in winning court cases — at which they have been notably unsuccessful.[21] Nor does it spring from the activities of the able debaters who fan out from the creationist institutions in San Diego on nationwide speaking tours. These highly visible and effective spokesmen fire up their devotees, recruit new supporters, and provide encouragement, information, and propaganda materials. As charismatic figures, they are necessary to the creationist cause but are not sufficient to guarantee its success.

The great strength of the creationist movement lies in the many small groups of dedicated, sincere, energetic, zealous believers that exist in communities in Iowa and all over the country. In supporting creationism and deprecating evolution, these groups work both formally and informally. Their formal lobbying efforts use methods that have been traditional in American political life for generations. The creationist activists go to public meetings and make their voices heard. They organize meetings of their own. They distribute literature. They buttonhole people. They push doorbells. They send letters. They call in to talk shows. They lobby legislators and other public officials. Such standard political techniques, applied in Iowa, started Jimmy Carter on his way to the presidency. They work equally well for the creationists.

Even more effective, perhaps, is informal, personal pressure applied by creationists, as the following two examples illustrate. In one Iowa community several years ago, the president of the school board told a newly appointed biology teacher that he had every right to teach evolution if he wished — but not in that district if he wanted to keep his job. The teacher took the hint, started looking for a new job, and left at the end of the year. His successor, a young woman, was told substantially the same thing. She is still in the district, unhappily *not* teaching evolution. In how many thousands of school districts across the nation do similar conditions prevail?[22]

Our second example concerns a March 1981 *Time* article on the creation-evolution controversy.[23] In preparation for the article, *Time*'s Chicago correspondent asked the writers to arrange an interview with an Iowa teacher who had been forbidden to teach evolution. Area science consultant George Magrane helped the writers locate a number of teachers in this position. But not one of them would agree to be interviewed. None would even consent to talk with the correspondent off the record. In a small town, there is no such thing as "off the record"; there are no secrets. Thereupon the correspondent went to Minnesota and met essentially the same reaction. In the end the *Time* article quoted Magrane: "Teachers in Iowa are being intimidated by the controversy. Rather than teach creationism and evolution, they teach neither one. It's almost a regression in history."

Scientists at the Grass Roots

Aware of the intimidation, Iowa scientists cast about during 1980–1981 for means of meeting creationism at the community or grass roots level. Two prototype strategies were developed. They were implemented through organizations called Committees of Correspondence (C/Cs), and the IAS Panel on Controversial Issues.[24]

The C/Cs originated in Iowa in the fall of 1980 and quickly became a nationwide movement. By August 1981 C/Cs were established in thirty-five states. Each C/C is a voluntary, independent, autonomous, proevolution group. The membership is composed primarily of scientists, but it also includes clergy, teachers, engineers, lawyers, physicians, members of other professions, and the general public. There is no national organization; the committees simply form a communications network. This pattern is by design. The intention is not to duplicate or to compete with the many existing proevolution organizations of national scope but rather to supplement them at state and local levels. The name "Committees of Correspondence" is evocative of the similarly named committees that formed in the thirteen colonies before the American Revolution. Leaders of the C/Cs are generally called "liaisons" to emphasize their nondirective, nonauthoritative, first among equals role. The stress is on local involvement and local leadership by many individuals.

The C/Cs engage in extensive public education activities and in a limited amount of lobbying. Their members write, sponsor, issue, and support publications. They develop proevolution publicity through the press and the electronic media. They testify before legislative committees, school boards, textbook committees, and other similar bodies. They hold and participate in meetings, debates, and symposia.

The success of the movement has been mixed. In 1980–1981, creationist bills were defeated in about twenty states. Testimony and letter-writing campaigns by C/C members helped in the defeat of a number of these bills. In two states the results of such efforts were lamentable. Arkansas passed an equal-time bill in March 1981, and Louisiana passed one in July 1981. In January 1982, in a suit brought by the American Civil Liberties Union and others, U.S. District Court Judge William R. Overton declared the Arkansas law unconstitutional in an impressive and far-reaching opinion, which is reprinted in Chapter 15.

The C/Cs have dissuaded local school districts from introducing creationism into their science curricula. But efforts at this level—the primary objective of the movement—are limited. Through frequent and close contact between the C/Cs of neighboring states, the committees hoped to improve their tactics and performance.

The IAS Panel on Controversial Issues is entirely an Iowa effort. The panel consists of about twenty-five scientists and several science educators, all IAS members. It sends teams of scientists to meet with school boards, school faculties, parent groups, and similar bodies in communities in which controversies exist over science issues. The current issue is mainly creation-evolution, but the panel is prepared to deal with other issues as well. The panel enters a community only upon invitation. It does not seek confrontation or legal sanctions. Its aim is to represent the position of the scientific community on the issue in question and to develop some understanding for scientific rationalism in a public that may have only a vague notion of the methods and goals of science.

The first invitations from school districts reached the IAS panel in a curious way. After the defeat of one of his creationist bills, State Senator John W. Jensen sent a letter on an official letterhead to school districts in the state (Figure 14.1). The letter asked the school districts, on the senator's initiative, to teach a creation-evolution two-model approach. The letter puzzled many of its recipients. Since it was on a senate letterhead, how official was the request? How should they respond? School officials in a number of districts addressed such inquiries to the DPI, which responded by sending them the DPI position paper.

As public documents, the inquiries from school districts were accessible to the IAS panel, which took them to be invitations to discussion and thus followed up on them. This action seems appropriate, as the IAS is a quasi-official body. It receives a small state subsidy and provides regular consultation services to various departments of the state govenment. Since the first flood of inquiries from school district officials, additional invitations have come to the panel through other channels. The panel began its work in the summer of 1981. Its experience is still too meager, however, to attempt to evaluate the effectiveness of its strategy.

Evolution is commonly taught in the public schools of the United States because the scientific community recognizes it as a valid and well-confirmed theory. Yet, as of July 1982, state law in Louisiana and school board decisions in several localities required creationism to be taught in courses that deal with the origin of human life, the earth, or the universe, including such courses as biology, life science, anthropology, sociology, physics, chemistry, world history, philosophy, and social studies.

The creationist argument seems to represent one direction along which the burgeoning religious right is showing its determination to mandate acceptance of its ideas in all aspects of our society. Citizens must recognize such actions for what they are.

JOHN W. JENSEN
STATE SENATOR
District 19

HOME ADDRESS
R.R Box 107
PLAINFIELD, IOWA, 50666

HOME PHONE
(319) 276-4445

COMMITTEES
Transportation, Vice Chair
Education
Labor & Industrial Relations
Claim-Appropriations Subcommittee

The Senate
State of Iowa
Sixty-Eighth General Assembly
State House
Des Moines, Iowa 50319

Dear Friend:

I believe that you have a concern for the best education you can offer the students of your school. I am concerned about the teaching of only model of the origin of man.

Your students would have a better concept of the subject if the scientific evidence supporting both concepts were taught.

I believe we are walking on dangerous ground according to some constitutional lawyers. Yale Law Journal has said that according to their opinion teaching either evolution or the creative model alone could be declared unconstitutional.

As far as the idea of religion The American Humanist associates proclaim "There are no alternatives to the principle of evolution". Humanism was declared a religion by the Supreme Court in 1976.

There are science books that have both models presented in them, or you could look for these text books or order some short course books on the subject. Two that I know of are Origins Two Models Evolution Creation by Richard B. Bliss presents both models and is aimed at the 5, 6, 7 and 8th grade level. Origin of Life, Evolution and Creation by Richard Bliss and Gary Parker could be used in junior high and high school. Both are published by Creation Life Publications, San Diego, California. These are only two books that I happen to have, the D.P.I. has a list of text books and support material that contains both models. Area 11 has film strips and other teaching helps in their library and I am sure other area education units can obtain such material.

In none of the science creation literature that I have read is the Bible ever quoted or used in the teaching of the creation model, only the scientific evidence supporting this model.

My concern has been and is to offer a balanced education to our students, as far as having religion in your science class, I am convinced that this is as much science in the creation model as the evolution model and as much religion in the evolution model as there is in the creation model.

I have a letter written by Dr. Robert Benton, March 9, 1978 pointing out that it is legal to present both models. "The paper creation, evolution and public education place no restriction on the Urbandale board of directors or any other public or private school in the state. The determination of the place for the "evolution or creation models" in the science curriculum is one of local determination".

So, the decision is yours. I feel balancing your teaching on this subject would be a step to help a fuller education of the students of your school.

If you wish any future information, get in touch with myself or the D.P.I. and the information will be forthcoming.

Thanks for your time.

John W. Jensen
State Senator

jwj/dc

FIG. 14.1. Letter from Iowa Senator John W. Jensen to school districts in Iowa.

APPENDIX A

State of Iowa
DEPARTMENT OF PUBLIC INSTRUCTION
Curriculum Division, Grimes State Office Building, Des Moines, Iowa

CREATION, EVOLUTION AND PUBLIC EDUCATION

THE POSITION OF THE
IOWA DEPARTMENT OF PUBLIC INSTRUCTION

THE CONTROVERSY

In Iowa and other states, "creationism" has recently been advanced as an alternative to the theory of evolution. Attempts have been made to legislatively mandate "equal time" for creationist concepts in science classrooms, materials and textbooks.

Interviews and surveys conducted by the Iowa Department of Public Instruction show that most Iowa religious leaders, science educators, scientists and philosophers contacted support the present patterns of teaching science in Iowa's schools. In addition, due to the nature of scientific and theological concepts, these authorities feel that the specifics of each discipline should be confined to their respective houses.

The National Academy of Science has stated that religion and science are "separate and mutually exclusive realms of human thought whose presentation in the same context leads to misunderstanding of both scientific theories and religious beliefs."[1]

CREATIONISM

In America, religion is usually defined as the expression of man's belief in, and reverence for, a metaphysical power governing all activities of the universe. Where there is not belief in metaphysical power, religion is a concern for that which is ultimate. Generally creationism is a religious concept. It proposes that all living things were created by a Creator. According to the creation model, "all living things originated from basic kinds of life, each of which was separately created."[2]

There are many versions of creation. Generally, creationists advocate that all permanent, basic life forms originated thousands of years ago through directive acts of a Creator—independent of the natural universe. Plants and animals were created separately with their full genetic potentiality provided by the Creator. Any variation, or speciation, which has occurred since creation has been within the original prescribed boundaries. Since each species contains its full potentiality, nature is viewed as static, reliable and predictable. Based on alleged gaps in the geologic record, creationists reject the theory of the descent of plants and animals from a single line of ancestors arising through random mutation and successively evolving over billions of years. It is further alleged that, through analysis of geologic strata, the earth has experienced at least one great flood or other natural global disasters accounting for the mass extinction of many biological organisms. Following such extinctions there followed sudden increases in the number, variety and complexity of organisms.

Having all Biblical accounts of creationism placed in comparative theology courses with other religious accounts of origins will not placate ardent creationists. They require

---

[1] Resolutions adopted by the National Academy of Science and the Commission of Science Education of the American Academy for the Advancement of Science (Washington, D.C.: October 17, 1972).

[2] R. B. Bliss, *Origins: Two Models; Evolution, Creation* (San Diego: Creation Life Publishers, 1976), p. 31.

that creationism be presented as a viable scientific alternative to evolution.[3] More zealous creationists argue that "it is only in the Bible that we can possibly obtain any information about the methods of creation, the order of creation, the duration of creation, or any other details of creation."[4]

SCIENCE

Science is an attempt to help explain the world of which we are a part. It is both an *investigatory process* and a *body of knowledge* readily subjected to investigation and verification. By a generally accepted definition, science is *not* an indoctrination process, but rather an objective method for problem solving. Science is an important part of the foundation upon which rest our technology, our agriculture, or economy, our intellectual life, our national defense, and our ventures into space.

The formulation of theories is a basic part of scientific method. Theories are generalizations, based on substantial evidence, which explain many diverse phenomena. A theory is always tentative. It is subject to test through the uncovering of new data, through new experiments, through repetition and refinements of old experiments, or through new interpretations. Should a significant body of contrary evidence appear, the theory is either revised or it is replaced by a new and better theory. The strength of a scientific theory lies in the fact that it is the most logical explanation of known facts, principles, and concepts dealing with an idea which does not currently have a conclusive test.

EVOLUTION

The *theory* of evolution meets the criteria of a scientific theory. It can explain much of the past and help predict many future scientific phenomena. Basically, the theory states that modern biologic organisms descended, with modification, from pre-existing forms which in turn had ancestors. Those organisms best adapted, through anatomical and physiological modification to their environment, left more offspring than did non-adapted organisms. The increased diversity of organisms enhanced their ability to survive in various environments and enabled them to leave more progeny.

The theory of evolution is designed to answer the "how" questions of science and biological development; it cannot deal effectively with the "who" or "why" of man's origin and development. It is, however, an effective means of integrating and clarifying many otherwise isolated scientific facts, principles and concepts.

There have been alternatives proposed to the theory of evolution (i.e., creationism, exo-biology, spontaneous generation); however, none are supported by the amount of scientific evidence that presently supports the theory of evolution.

It is evident that the *process* of evolution occurs. Successful species of living organisms change with time when exposed to environmental pressures. Such changes in species have been documented in the past, and it can be confidently predicted that they will continue to change in the future. Evolution helps explain many other scientific phenomena: variations in disease, drug resistance in microbes, anatomical anomalies which appear in surgery, and successsful methods for breeding better crops and farm animals. Modern biological science and its applications on the farm, in medicine, and elsewhere are not completely understandable without many of the basic concepts of evolution.

There are many things that evolution is not. It is not dogma. Although there is intense dispute among scientists concerning the details of evolution, most scientists accept its validity on the ground of its strong supporting evidence.

---

[3]Morris, Henry M., *The Remarkable Birth of Planet Earth* (San Diego: Creation Life Publishers, 1972).

[4]National Association of Biology Teachers—A Compendium of Information on the Theory of Evolution and the Evolution-Creationism Controversy (June 1977).

DEPARTMENT OF PUBLIC INSTRUCTION DECISION

Teaching religious doctrine is not the science teacher's responsibility. Teachers should recognize the personal validity of alternative beliefs, but should then direct student inquiries to the appropriate institution for counseling and/or further explanation. Giving equal emphasis in science classes to non-scientific theories that are presented as alternatives to evolution would be in direct opposition to understanding the nature and purpose of science.

Each group is fully entitled to its point of view with respect to the Bible and evolution; but the American doctrine of religious freedom and the Establishment Clause in the First Amendment to the U.S. Constitution forbid either group—or any other religious group—from pressing its point of view on the public schools. An Indiana court decision declared: "The prospect of biology teachers and students alike forced to answer and respond to continued demand for 'correct' Fundamentalist Christian doctrines has no place in public schools."[5]

The science curriculum should emphasize the theory of evolution as a well-supported scientific theory—not a fact—that is taught as such by certificated science teachers. Students should be advised that it is their responsibility, as informed citizens, to have creationism explained to them by theological experts. They must then decide for themselves the merits of each discipline and its relevance to their lives.

The Iowa Department of Public Instruction feels that public schools cannot be surrogate family, church and all other necessary social institutions for students, and for them to attempt to do so would be a great disservice to citizens and appropriate institutions.

APPENDIX B

STATEMENT OF THE POSITION OF THE IOWA ACADEMY OF
SCIENCE ON THE STATUS OF CREATIONISM AS A
SCIENTIFIC EXPLANATION OF NATURAL PHENOMENA
31 January 1981

Current attempts to introduce "scientific creationism" into the science classroom are strongly opposed by The Iowa Academy of Science on the grounds that creationism when called "scientific" is a religious doctrine posed as science. It is contrary to the nature of science to propose supernatural explanations of natural events or their origins. With its appeal to the supernatural, creationism is outside the realm of science.

Creationist organizations that are advocating the teaching of "scientific creationism" in science classrooms include members purported to be scientists who have examined the evidence and have found creationism to be a superior alternative to evolution. They claim to know of evidence that supports the idea of a young earth and that shows evolution to be impossible. Much of this "evidence" is inaccurate, out of date, and not accepted by recognized paleontologists and biologists. The total membership of these "scientific" creationist groups constitutes only a fraction of one percent of the scientific personnel in this country. Most of them are not trained in biology or geology, the areas in which professional judgments are made in the field of evolutionary theory. They often misrepresent the positions of respected scientists and quote them out of context to support their own views before audiences and government bodies. They are driven by the notion that all explanations of natural events must conform to their preconceived creationist views. These tactics are used to give the uninformed public the false impression that science itself is confused. Then a supernatural explanation is proposed to bring order out of apparent chaos.

The Iowa Academy of Science urges legislators, school administrators, and the general public not to be misled by the tactics of these so-called "scientific creationists."

---

[5]Hendren vs. Campbell, Supreme Court No. 5 Marion County, Indiana (1977), p. 20.

The Academy respects the right of persons to hold diverse religious beliefs, including those who reject evolution, but only as matters of theology or faith, not as secular science. Creationism is not science and the Academy deplores and opposes any attempt to disguise it as science. Most recognized scientists find no conflict between religious faith and acceptance of evolution. They do not view evolution as being anti-religious. They have no vested interest in supporting evolution as do the "scientific creationists" in supporting creationism, but merely consider evolution as being most consistent with the best evidence.

The Iowa Academy of Science feels strongly that the distinction between science and religion must be maintained. A state with one of the highest literacy rates and with the highest scientific literacy scores in the nation, and one which prides itself on the individuality of its citizens, should discriminate in its pubic education system between what is science and what is not science.

(Approved by a majority of all voting members of the Iowa Academy of Science in February, 1981)

APPENDIX C

POSITION STATEMENT ON THE CREATION/EVOLUTION
CONTROVERSY FROM THE IOWA COUNCIL
OF SCIENCE SUPERVISORS (CS²)

Because of the insistence that special creation be taught in Iowa science classes as an alternative concept to evolution, we, the Iowa Council of Science Supervisors, as representatives of the science educators in Iowa, make the following statement:

Science educators are responsible for interpreting the spirit and substance of science to their students. Teachers are bound to promote a scientific rationale based upon carefully defined and objective judgments of scientific endeavors. When conflicts arise between competing paradigms in science, they must be resolved by the scientific communtiy rather than by the educators of science.

Based upon court decisions in Indiana and Tennessee, and in the creationists' own statements of beliefs, the Creation Research Society is premised upon the full belief in the Biblical record of special creation.

> The Bible is the Written Word of God, and because it is inspired throughout, all its assertions are historically and scientifically true in all original autographs. To the student of nature this means that the account of origins in Genesis is a factual presentation of simple historical truth.*

Science is tentative and denies an ultimate or perfect truth as claimed by scientific creationism. We suggest that creationists submit their creation theories and models to recognized science organizations such as the American Association for the Advancement of Science (AAAS) or their affiliated scientific societies. The claims of these paradigms should be substantiated with validated objective evidence. The scientific organizations would assume responsibility for analyzing the materials, making their findings available for national review through AAAS scientific journals.

Until "scientific creation" receives substantial support from such organizations as AAAS, American Anthropological Association, State Academies of Science, National Academy of Science, and National Paleontological and Geological Associations, it is recommended that this organization (CS²) and the science teachers of Iowa reject further consideration of scientific creationism as an alternative approach to established science teaching practices.

---

*Membership application forms for the Creation Research Society, Wilbert II. [*sic,* should be H.] Rusch, Membership Secretary, 2712 Cranbrook Road, Ann Arbor, Michigan 48104.

[15]

The Battle in Arkansas:
The Judge's Decision

EDITORS' NOTE: In Arkansas in March 1981, Act 590 of 1981, entitled the Balanced Treatment for Creation-Science and Evolution-Science Act became law. In May 1981 a suit was filed challenging it. The subsequent trial in December 1981 led to Judge William R. Overton's decision in January 1982, most of which we publish here. The complete text can be found in "Creationism in the Schools: The Arkansas Decision," *American Biology Teacher* 44(March 1982):172–79 and in "Creationism in Schools: The Decision in McLean versus the Arkansas Board of Education," *Science* 215(February 1982):934–43. Judge Overton found in favor of the plaintiffs. Chapter 13 discusses the decision. For accounts of the trial, see the references cited in note 4 of the Introduction. In this chapter we pick up the decision with Judge Overton's discussion of present-day creationism. We have omitted the decision's footnotes, some of its legal references, and its first few pages, which cover background material found in Chapters 2 and 13.

IN the 1960's and early 1970's, several Fundamentalist organizations were formed to promote the idea that the Book of Genesis was supported by scientific data. The terms "creation science" and "scientific creationism" have been adopted by these Fundamentalists as descriptive of their study of creation and the origins of man. Perhaps the leading creationist organization is the Institute for Creation Research (ICR), which is affiliated with the Christian Heritage College and supported by the Scott Memorial Baptist Church in San Diego, California. The ICR, through the Creation-Life Publishing Company, is the leading publisher of creation science material. Other creation science organizations include the Creation Science Research Center (CSRC) of San Diego and the Bible Science Association of Minneapolis, Minnesota. In 1963, the Creation Research Society (CRS) was formed from a schism in the American Scientific Affiliation (ASA). It is an organization of literal Fundamentalists who have the equivalent of a master's degree in some recognized area of science. A purpose of the organization is "to reach all people with the vital message of the scientific and historic truth about creation." Similarly, the CSRC was formed in 1970 from a split in the CRS. Its aim has been "to reach the 63 million children of the United States with the scientific teaching of Biblical creationism."

Among creationist writers who are recognized as authorities in the field by other creationists are Henry M. Morris, Duane Gish, G. E. Parker, Harold S. Slusher, Richard B. Bliss, John W. Moore, Martin E. Clark, W. L. Wysong,

Robert E. Kofahl and Kelly L. Segraves. Morris is Director of ICR, Gish is Associate Director and Segraves is associated with CSRC.

Creationists view evolution as a source of society's ills, and the writings of Morris and Clark are typical expressions of that view.

> Evolution is thus not only anti-Biblical and anti-Christian, but it is utterly unscientific and impossible as well. But it has served effectively as the pseudo-scientific basis of atheism, agnosticism, socialism, fascism, and numerous other false and dangerous philosophies over the past century.

Creationists have adopted the view of Fundamentalists generally that there are only two positions with respect to the origins of the earth and life: belief in the inerrancy of the Genesis story of creation and of a worldwide flood as fact, or belief in what they call evolution.

Henry Morris has stated, "It is impossible to devise a legitimate means of harmonizing the Bible with evolution." This dualistic approach to the subject of origins permeates the creationist literature.

The creationist organizations consider the introduction of creation science into the public schools part of their ministry. The ICR has published at least two pamphlets containing suggested methods for convincing school boards, administrators and teachers that creationism should be taught in public schools. The ICR has urged its proponents to encourage school officials to voluntarily add creationism to the curriculum.

Citizens For Fairness In Education is an organization based in Anderson, South Carolina, formed by Paul Ellwanger, a respiratory therapist who is trained in neither law nor science. Mr. Ellwanger is of the opinion that evolution is the forerunner of many social ills, including Nazism, racism and abortion. About 1977, Ellwanger collected several proposed legislative acts with the idea of preparing a model state act requiring the teaching of creationism as science in opposition to evolution. One of the proposals he collected was prepared by Wendell Bird, who is now a staff attorney for ICR. From these various proposals, Ellwanger prepared a "model act" which calls for "balanced treatment" of "scientific creationism" and "evolution" in public schools. He circulated the proposed act to various people and organizations around the country.

Mr. Ellwanger's views on the nature of creation science are entitled to some weight since he personally drafted the model act which became Act 590. His evidentiary deposition with exhibits and unnumbered attachments (produced in response to a subpoena *duces tecum*) speaks to both the intent of the Act and the scientific merits of creation science. Mr. Ellwanger does not believe creation science is a science. In a letter to Pastor Robert E. Hays he states, "While neither evolution nor creation can qualify as a scientific theory, and since it is virtually impossible at this point to educate the whole world that evolution is not a true scientific theory, we have freely used these terms—the evolution theory and the theory of scientific creationism—in the bill's text." He further states in a letter to Mr. Tom Bethell, "As we examine evolution (remember, we're not making any scientific claims for creation, but we are challenging evolution's claim to be scientific). . . ."

Ellwanger's correspondence on the subject shows an awareness that Act 590 is a religious crusade, coupled with a desire to conceal this fact. In a letter to State Senator Bill Keith of Louisiana, he says, "I view this whole battle as one between God and anti-God forces, though I know there are a large number of evolutionists who believe in God." And further, ". . . it behooves Satan to do all he can to thwart our efforts and confuse the issue at every turn." Yet Ellwanger suggests to Senator Keith, "If you have a clear choice between having grassroots leaders of this statewide bill promotion effort to be ministerial or non-ministerial, be sure to opt for the non-ministerial. It does the bill effort no good to have ministers out there in the public forum and the adversary will surely pick at this point. . . . Ministerial persons can accomplish a tremendous amount of work from behind the scenes, encouraging their congregations to take the organizational and P.R. initiatives. And they can lead their churches in storming Heaven with prayers for help against so tenacious an adversary."

Ellwanger shows a remarkable degree of political candor, if not finesse, in a letter to State Senator Joseph Carlucci of Florida:

> 2. It would be very wise, if not actually essential, that all of us who are engaged in this legislative effort be careful not to present our position and our work in a religious framework. For example, in written communications that might somehow be shared with those other persons whom we may be trying to convince, it would be well to exclude our own personal testimony and/or witness for Christ, but rather, if we are so moved, to give that testimony on a separate attached note.

The same tenor is reflected in a letter by Ellwanger to Mary Ann Miller, a member of FLAG (Family, Life, America under God) who lobbied the Arkansas Legislature in favor of Act 590:

> . . . we'd like to suggest that you and your co-workers be very cautious about mixing creation-science with creation-religion. . . . Please urge your co-workers not to allow themselves to get sucked into the "religion" trap of mixing the two together, for such mixing does incalculable harm to the legislative thrust. It could even bring public opinion to bear adversely upon the higher courts that will eventually have to pass judgment on the constitutionality of this new law.

Perhaps most interesting, however, is Mr. Ellwanger's testimony in his deposition as to his strategy for having the model act implemented:

> Q. You're trying to play on other people's religious motives.
>
> A. I'm trying to play on their emotions, love, hate, their likes, dislikes, because I don't know any other way to involve, to get humans to become involved in human endeavors. I see emotions as being a healthy and legitimate means of getting people's feelings into action, and . . . I believe that the predominance of population in America that represents the greatest potential for taking some kind of action in this area is a Christian community. I see the Jewish community as far less poten-

tial in taking action . . . but I've seen a lot of interest among Christians and I feel, why not exploit that to get the bill going if that's what it takes.

Mr. Ellwanger's ultimate purpose is revealed in the closing of his letter to Mr. Tom Bethell: "Perhaps all this is old hat to you, Tom, and if so, I'd appreciate your telling me so and perhaps where you've heard it before — the idea of killing evolution instead of playing these debating games that we've been playing for nigh over a decade already."

It was out of this milieu that Act 590 emerged. The Reverend W. A. Blount, a Biblical literalist who is pastor of a church in the Little Rock area and was, in February, 1981, chairman of the Greater Little Rock Evangelical Fellowship, was among those who received a copy of the model act from Ellwanger.

At Reverend Blount's request, the Evangelical Fellowship unanimously adopted a resolution to seek introduction of Ellwanger's act in the Arkansas Legislature. A committee composed of two ministers, Curtis Thomas and W. A. Young, was appointed to implement the resolution. Thomas obtained from Ellwanger a revised copy of the model act which he transmitted to Carl Hunt, a business associate of Senator James L. Holsted, with the request that Hunt prevail upon Holsted to introduce the act.

Holsted, a self-described "born again" Christian Fundamentalist, introduced the act in the Arkansas Senate. He did not consult the State Department of Education, scientists, science educators or the Arkansas Attorney General. The Act was not referred to any Senate committee for hearing and was passed after only a few minutes' discussion on the Senate floor. In the House of Representatives, the bill was referred to the Education Committee which conducted a perfunctory fifteen minute hearing. No scientist testified at the hearing, nor was any representative from the State Department of Education called to testify.

Ellwanger's model act was enacted into law in Arkansas as Act 590 without amendment or modification other than minor typographical changes. The legislative "findings of fact" in Ellwanger's act and Act 590 are identical, although no meaningful fact-finding process was employed by the General Assembly.

Ellwanger's efforts in preparation of the model act and campaign for its adoption in the states were motivated by his opposition to the theory of evolution and his desire to see the Biblical version of creation taught in the public schools. There is no evidence that the pastors, Blount, Thomas, Young or The Greater Little Rock Evangelical Fellowship were motivated by anything other than their religious convictions when proposing its adoption or during their lobbying efforts in its behalf. Senator Holsted's sponsorship and lobbying efforts in behalf of the Act were motivated solely by his religious beliefs and desire to see the Biblical version of creation taught in the public schools.

The State of Arkansas, like a number of states whose citizens have relatively homogeneous religious beliefs, has a long history of official opposition to evolution which is motivated by adherence to Fundamentalist beliefs in the inerrancy of the Book of Genesis. This history is documented in Justice Fortas' opinion in *Epperson v. Arkansas,* 393 U.S. 97 (1968), which struck down Initiated Act 1 of 1929, Ark. Stat. Ann. §§80-1627-1628, prohibiting the teaching of the theory of evolution. To this same tradition may be attributed Initiated Act 1 of 1930, Ark. Stat. Ann. §80-1606 (Repl. 1980), requiring "the reverent daily reading of a portion of the English Bible" in every public school classroom in the state.

It is true, as defendants argue, that courts should look to legislative statements of a statute's purpose in Establishment Clause cases and accord such pronouncements great deference. Defendants also correctly state the principle that remarks by the sponsor or author of a bill are not considered controlling in analyzing legislative intent.

Courts are not bound, however, by legislative statements of purpose or legislative disclaimers. In determining the legislative purpose of a statute, courts may consider evidence of the historical context of the Act, the specific sequence of events leading up to passage of the Act, departures from normal procedural sequences, substantive departures from the normal, and contemporaneous statements of the legislative sponsor. . . .

The unusual circumstances surrounding the passage of Act 590, as well as the substantive law of the First Amendment, warrant an inquiry into the stated legislative purposes. The author of the Act had publicly proclaimed the sectarian purpose of the proposal. The Arkansas residents who sought legislative sponsorship of the bill did so for a purely sectarian purpose. These circumstances alone may not be particularly persuasive, but when considered with the publicly announced motives of the legislative sponsor made contemporaneously with the legislative process; the lack of any legislative investigation, debate or consultation with any educators or scientists; the unprecedented intrusion in school curriculum; and official history of the State of Arkansas on the subject, it is obvious that the statement of purposes has little, if any, support in fact. The State failed to produce any evidence which would warrant an inference or conclusion that at any point in the process anyone considered the legitimate educational value of the Act. It was simply and purely an effort to introduce the Biblical version of creation into the public school curricula. The only inference which can be drawn from these circumstances is that the Act was passed with the specific purpose by the general Assembly of advancing religion. The Act therefore fails the first prong of the three-pronged test, that of secular legislative purpose, as articulated in *Lemon v. Kurtzman* and *Stone v. Graham.*

III.

If the defendants are correct and the Court is limited to an examination of the language of the Act, the evidence is overwhelming that both the purpose and effect of Act 590 is the advancement of religion in the public schools.

Section 4 of the Act provides:

Definitions. As used in this Act:

(a) "Creation-science" means the scientific evidences for creation and inferences from those scientific evidences. Creation-science includes the scientific evidences and related inferences that indicate: (1) Sudden creation of the universe, energy, and life from nothing; (2) The insufficiency of mutation and natural selection in bringing about development of all living kinds from a single organism; (3) Changes only within fixed limits of originally created kinds of plants and animals; (4) Separate ancestry for man and apes; (5) Explanation of the earth's geology by catastrophism, including the occurrence of a worldwide flood; and (6) A relatively recent inception of the earth and living kinds.

(b) "Evolution-science" means the scientific evidences for evolution and inferences from those scientific evidences. Evolution-science includes the scientific evidences and related inferences that indicate: (1) Emergence by naturalistic processes of the universe from disordered matter and emergence of life from nonlife; (2) The sufficiency of mutation and natural selection in bringing about development of present living kinds from simple earlier kinds; (3) Emergence by mutation and natural selection of present living kinds from simple earlier kinds; (4) Emergence of man from a common ancestor with apes; (5) Explanation of the earth's geology and the evolutionary sequence by uniformitarianism; and (6) An inception several billion years ago of the earth and somewhat later of life.

(c) "Public schools" mean public secondary and elementary schools.

The evidence establishes that the definition of "creation science" contained in 4(a) has as its unmentioned reference the first 11 chapters of the Book of Genesis. Among the many creation epics in human history, the account of sudden creation from nothing, or *creatio ex nihilo,* and subsequent destruction of the world by flood is unique to Genesis. The concepts of 4(a) are the literal Fundamentalists' view of Genesis. Section 4(a) is unquestionably a statement of religion, with the exception of 4(a)(2) which is a negative thrust aimed at what the creationists understand to be the theory of evolution.

Both the concepts and wording of Section 4(a) convey an inescapable religiosity. Section 4(a)(1) describes "sudden creation of the universe, energy and life from nothing." Every theologian who testified, including defense witnesses, expressed the opinion that the statement referred to a supernatural creation which was performed by God.

Defendants argue that: (1) the fact that 4(a) conveys ideas similar to the literal interpretation of Genesis does not make it conclusively a statement of religion; (2) that reference to a creation from nothing is not necessarily a religious concept since the Act only suggests a creator who has power, intelligence and a sense of design and not necessarily the attributes of love, compassion and justice; and (3) that simply teaching about the concept of a creator is not a religious exercise unless the student is required to make a commitment to the concept of a creator.

The evidence fully answers these arguments. The ideas of 4(a)(1) are not merely similar to the literal interpretation of Genesis; they are identical and parallel to no other story of creation.

The argument that creation from nothing in 4(a)(1) does not involve a supernatural deity has no evidentiary or rational support. To the contrary, "creation out of nothing" is a concept unique to Western religions. In traditional

Western religious thought, the conception of a creator of the world is a conception of God, Indeed, creation of the world "out of nothing" is the ultimate religious statement because God is the only actor. As Dr. Langdon Gilkey noted, the Act refers to one who has the power to bring all the universe into existence from nothing. The only "one" who has this power is God.

The leading creationist writers, Morris and Gish, acknowledge that the idea of creation described in 4(a)(1) is the concept of creation by God and make no pretense to the contrary. The idea of sudden creation from nothing, or *creatio ex nihilo,* is an inherently religious concept.

The argument advanced by defendants' witness, Dr. Norman Geisler, that teaching the existence of God is not religious unless the teaching seeks a commitment, is contrary to common understanding and contradicts settled case law.

The facts that creation science is inspired by the Book of Genesis and that Section 4(a) is consistent with a literal interpretation of Genesis leave no doubt that a major effect of the Act is the advancement of particular religious beliefs. The legal impact of this conclusion will be discussed further at the conclusion of the Court's evaluation of the scientific merit of creation science.

IV.(A)

The approach to teaching "creation science" and "evolution science" found in Act 590 is identical to the two-model approach espoused by the Institute for Creation Research and is taken almost verbatim from ICR writings. It is an extension of Fundamentalists' view that one must either accept the literal interpretation of Genesis or else believe in the godless system of evolution.

The two-model approach of the creationists is simply a contrived dualism which has no scientific factual basis or legitimate educational purpose. It assumes only two explanations for the origins of life and existence of man, plants and animals: It was either the work of a creator or it was not. Application of these two models, according to creationists and the defendants, dictates that all scientific evidence which fails to support the theory of evolution is necessarily scientific evidence in support of creationism and is, therefore, creation science "evidence" in support of Section 4(a).

IV.(B)

The emphasis on origins as an aspect of the theory of evolution is peculiar to creationist literature. Although the subject of origins of life is within the province of biology, the scientific community does not consider origins of life a part of evolutionary theory.* The theory of evolution assumes the existence of life and is directed to an explanation of *how* life evolved. Evolution does not presuppose the absence of a creator or God and the plain inference conveyed by Section 4 is erroneous.

[*This is true for biological evolution, which the trial focused on. However, in the creation-evolution controversy, "evolutionary theory" has been construed broadly enough to include the origin of life and much else.]

As a statement of the theory of evolution, Section 4(b) is simply a hodge-podge of limited assertions, many of which are factually inaccurate.

For example, although 4(b)(2) asserts, as a tenet of evolutionary theory, "the sufficiency of mutation and natural selection in bringing about the existence of present living kinds from simple earlier kinds," Drs. [Francisco J.] Ayala and [Stephen Jay] Gould both stated that biologists know that these two processes do not account for all significant evolutionary change. They testified to such phenomena as recombination, the founder effect, genetic drift and the theory of punctuated equilibrium, which are believed to play important evolutionary roles. Section 4(b) omits any reference to these. Moreover, 4(b) utilizes the term "kinds" which all scientists said is not a word of science and has no fixed meaning. Additionally, the Act presents both evolution and creation science as "package deals." Thus, evidence critical of some aspect of what the creationists define as evolution is taken as support for a theory which includes a worldwide flood and a relatively young earth.

IV.(C)

In addition to the fallacious pedagogy of the two-model approach, Section 4(a) lacks legitimate educational value because "creation science" as defined in that section is simply not science. Several witnesses suggested definitions of science. A descriptive definition was said to be that science is what is "accepted by the scientific community" and is "what scientists do." The obvious implication of this description is that, in a free society, knowledge does not require the imprimatur of legislation in order to become science.

More precisely, the essential characteristics of science are:

(1) It is guided by natural law;
(2) It has to be explanatory by reference to natural law;
(3) It is testable against the empirical world;
(4) Its conclusions are tentative, i.e., are not necessarily the final word; and
(5) It is falsifiable.

Creation science as described in Section 4(a) fails to meet these essential characteristics. First, the section revolves around 4(a)(1) which asserts a sudden creation "from nothing." Such a concept is not science because it depends upon a supernatural intervention which is not guided by natural law. It is not explanatory by reference to natural law, is not testable and is not falsifiable.

If the unifying idea of supernatural creation by God is removed from Section 4, the remaining parts of the section explain nothing and are meaningless assertions.

Section 4(a)(2), relating to the "insufficiency of mutation and natural selection in bringing about development of all living kinds from a single organism", is an incomplete negative generalization directed at the theory of evolution.

Section 4(a)(3) which describes "changes only within fixed limits of originally created kinds of plants and animals" fails to conform to the essential characteristics of science for several reasons. First, there is no scientific definition of "kinds"and none of the witnesses was able to point to any scientific authority which recognized the term or knew how many "kinds" existed. One defense witness suggested there may be 100 to 10,000 different "kinds". Another believes there were "about 10,000, give or take a few thousand." Second, the assertion appears to be an effort to establish outer limits of changes within species. There is no scientific explanation for these limits which is guided by natural law and the limitations, whatever they are, cannot be explained by natural law.

The statement in 4(a)(4) of "separate ancestry of man and apes" is a bald assertion. It explains nothing and refers to no scientific fact or theory.

Section 4(a)(5) refers to "explanation of the earth's geology by catastrophism, including the occurrence of a worldwide flood." This assertion completely fails as science. The Act is referring to the Noachian flood described in the Book of Genesis. The creationist writers concede that *any* kind of Genesis Flood depends upon supernatural intervention. A worldwide flood as an explanation of the world's geology is not the product of natural law, nor can its occurrence be explained by natural law.

Section 4(a)(6) equally fails to meet the standards of science. "Relatively recent inception" has no scientific meaning. It can only be given meaning by reference to creationist writings which place the age at between 6,000 and 20,000 years because of the genealogy of the Old Testament. Such a reasoning process is not the product of natural law; not explainable by natural law; nor is it tentative.

Creation science, as defined in Section 4(a), not only fails to follow the canons defining scientific theory, it also fails to fit the more general descriptions of "what scientists think" and "what scientists do." The scientific community consists of individuals and groups, nationally and internationally, who work independently in such varied fields as biology, paleontology, geology and astronomy. Their work is published and subject to review and testing by their peers. The journals for publication are both numerous and varied. There is, however, not one recognized scientific journal which has published an article espousing the creation science theory described in Section 4(a). Some of the State's witnesses suggested that the scientific community was "close-minded" on the subject of creationism and that explained the lack of acceptance of the creation science arguments. Yet no witness produced a scientific article for which publication had been refused. Perhaps some members of the scientific community are resistant to new ideas. It is, however, inconceivable that such a loose knit group of independent thinkers in all the varied fields of science could, or would, so effectively censor new scientific thought.

The creationists have difficulty maintaining among their ranks consistency in the claim that creationism is science. The author of Act 590, Ellwanger, said that neither evolution or creationism was science. He thinks both are religion. Duane Gish recently responded to an article in *Discover* critical of creationism by stating:

Stephen Jay Gould states that creationists claim creation is a scientific theory. This is a false accusation. Creationists have repeatedly stated that neither creation nor evolution is a scientific theory (and each is equally religious).

The methodology employed by creationists is another factor which is indicative that their work is not science. A scientific theory must be tentative and always subject to revision or abandonment in light of facts that are inconsistent with, or falsify, the theory. A theory that is by its own terms dogmatic, absolutist and never subject to revision is not a scientific theory.

The creationists' methods do not take data, weigh it against the opposing scientific data, and thereafter reach the conclusions stated in Section 4(a). Instead, they take the literal wording of the Book of Genesis and attempt to find scientific support for it. The method is best explained in the language of Morris in his book *Studies in The Bible and Science* at page 114:

> . . . it is . . . quite impossible to determine anything about Creation through a study of present processes, because present processes are not creative in character. If man wishes to know anything about Creation (the time of Creation, the duration of Creation, the order of Creation, the methods of Creation, or anything else) his sole source of true information is that of divine revelation. God was there when it happened. We were not there. . . . Therefore, we are completely limited to what God has seen fit to tell us, and this information is in His written Word. This is our textbook on the science of Creation!

The Creation Research Society employs the same unscientific approach to the issue of creationism. Its applicants for membership must subscribe to the belief that the Book of Genesis is "historically and scientifically true in all of the original autographs." The Court would never criticize or discredit any person's testimony based on his or her religious beliefs. While anybody is free to approach a scientific inquiry in any fashion they choose, they cannot properly describe the methodology used as scientific, if they start with a conclusion and refuse to change it regardless of the evidence developed during the course of the investigation.

IV.(D)

In efforts to establish "evidence" in support of creation science, the defendants relied upon the same false premise as the two-model approach contained in Section 4, i.e., all evidence which criticized evolutionary theory was proof in support of creation science. For example, the defendants established that the mathematical probability of a chance chemical combination resulting in life from non-life is so remote that such an occurrence is almost beyond imagination. Those mathematical facts, the defendants argue, are scientific evidences that life was the product of a creator. While the statistical figures may be impressive evidence against the theory of chance chemical combinations as an explanation of origins, it requires a leap of faith to interpret those figures so as to support a complex doctrine which includes a sudden creation from nothing, a worldwide flood, separate ancestry of man and apes, and a young earth.

The defendants' argument would be more persuasive if, in fact, there were only two theories or ideas about the origins of life and the world. That there are a number of theories was acknowledged by the State's witnesses, Dr. Wickramasinghe and Dr. Geisler. Dr. Wickramasinghe testified at length in support of a theory that life on earth was "seeded" by comets which delivered genetic material and perhaps organisms to the earth's surface from interstellar dust far outside the solar system. The "seeding" theory further hypothesizes that the earth remains under the continuing influence of genetic material from space which continues to affect life. While Wickramasinghe's theory about the origins of life on earth has not received general acceptance within the scientific community, he has, at least, used scientific methodology to produce a theory of origins which meets the essential characteristics of science.

The Court is at a loss to understand why Dr. Wickramasinghe was called in behalf of the defendants. Perhaps it was because he was generally critical of the theory of evolution and the scientific community, a tactic consistent with the strategy of the defense. Unfortunately for the defense, he demonstrated that the simplistic approach of the two-model analysis of the origins of life is false. Furthermore, he corroborated the plaintiffs' witnesses by concluding that "no rational scientist" would believe the earth's geology could be explained by reference to a worldwide flood or that the earth was less than one million years old.

The proof in support of creation science consisted almost entirely of efforts to discredit the theory of evolution through a rehash of data and theories which have been before the scientific community for decades. The arguments asserted by creationists are not based upon new scientific evidence or laboratory data which has been ignored by the scientific community.

Robert Gentry's discovery of radioactive polonium haloes in granite and coalified woods is, perhaps, the most recent scientific work which the creationists use as argument for a "relatively recent inception" of the earth and a "worldwide flood." The existence of polonium haloes in granite and coalified wood is thought to be inconsistent with radiometric dating methods based upon constant radioactive decay rates. Mr. Gentry's findings were published almost ten years ago and have been the subject of some discussion in the scientific community. The discoveries have not, however, led to the formulation of any scientific hypothesis or theory which would explain a relatively recent inception of the earth or a worldwide flood. Gentry's discovery has been treated as a minor mystery which will eventually be explained. It may deserve further investigation, but the National Science Foundation has not deemed it to be of sufficient import to support further funding.

The testimony of Marianne Wilson was persuasive evidence that creation science is not science. Ms. Wilson is in charge of the science curriculum for Pulaski County Special School District, the largest school district in the State of Arkansas. Prior to the passage of Act 590, Larry Fisher, a science teacher in the District, using materials from the ICR, convinced the School Board that it should voluntarily adopt creation science as part of its science curriculum. The District Superintendent assigned Ms. Wilson the job of producing a creation science curriculum guide. Ms. Wilson's testimony about the project was particularly convincing because she obviously approached the assignment with an open mind and no preconceived notions about the subject. She had not heard of creation science until about a year ago and did not know its meaning before she began her research.

Ms. Wilson worked with a committee of science teachers appointed from the District. They reviewed practically all of the creationist literature. Ms. Wilson and the committee members reached the unanimous conclusion that creationism is not science; it is religion. They so reported to the Board. The Board ignored the recommendation and insisted that a curriculum guide be prepared.

In researching the subject, Ms. Wilson sought the assistance of Mr. Fisher who initiated the Board action and asked professors in the science departments of the University of Arkansas at Little Rock and the University of Central Arkansas for reference material and assistance, and attended a workshop conducted at Central Baptist College by Dr. Richard Bliss of the ICR staff. Act 590 became law during the course of her work so she used Section 4(a) as a format for her curriculum guide.

Ms. Wilson found all available creationists' materials unacceptable because they were permeated with religious references and reliance upon religious beliefs.

It is easy to understand why Ms. Wilson and other educators find the creationists' textbook material and teaching guides unacceptable. The materials misstate the theory of evolution in the same fashion as Section 4(b) of the Act, with emphasis on the alternative mutually exclusive nature of creationism and evolution. Students are constantly encouraged to compare and make a choice between the two models, and the material is not presented in an accurate manner.

A typical example is *Origins* by Richard B. Bliss, Director of Curriculum Development of the ICR. The presentation begins with a chart describing "preconceived ideas about origins" which suggests that some people believe that evolution is atheistic. Concepts of evolution, such as "adaptive radiation," are erroneously presented. At page 11, figure 1.6, of the text, a chart purports to illustrate this "very important" part of the evolution model. The chart conveys the idea that such diverse mammals as a whale, bear, bat and monkey all evolved from a shrew through the process of adaptive radiation. Such a suggestion is, of course, a totally erroneous and misleading application of the theory. Even more objectionable, especially when viewed in light of the emphasis on asking the student to elect one of the models, is the chart presentation at page 17, figure 1.6. That chart purports to illustrate the evolutionists' belief that man evolved from bacteria to fish to reptile to mammals and, thereafter, into man. The illustration indicates, however, that the mammal from which man evolved was *a rat*.

Biology, A Search For Order in Complexity is a high school biology text typical of creationists' materials. The following quotations are illustrative:

Flowers and roots do not have a mind to have purpose of their own; therefore, this planning must have been done for them by the Creator.

The exquisite beauty of color and shape in flowers exceeds the skill of poet, artist, and king. Jesus said (from Matthew's gospel), "Consider the lilies of the field, how they grow; they toil not, neither do they spin."

The "public school edition" texts written by creationists simply omit Biblical references but the content and message remain the same. For example, *Evolution — The Fossils Say No!*, contains the following:

Creation. By creation we mean the bringing into being by a supernatural Creator of the basic kinds of plants and animals by the process of sudden, or fiat, creation.

We do not know how the Creator created, what processes He used, *for He used processes which are not now operating anywhere in the natural universe.* This is why we refer to creation as Special Creation. We cannot discover by scientific investigation anything about the creative processes used by the Creator.

Gish's book also portrays the large majority of evolutionists as "materialistic atheists or agnostics."

Scientific Creationism (Public School Edition) by Morris, is another text reviewed by Ms. Wilson's committee and rejected as unacceptable. The following quotes illustrate the purpose and theme of the text:

Parents and youth leaders today, and even many scientists and educators, have become concerned about the prevalence and influence of evolutionary philosophy in modern curriculum. Not only is this system inimical to orthodox Christianity and Judaism, but also, as many are convinced, to a healthy society and true science as well.

*   *   *   *   *

The rationalist of course finds the concept of special creation insufferably naive, even "incredible." Such a judgment, however, is warranted only if one categorically dismisses the existence of an omnipotent God.

Without using creationist literature, Ms. Wilson was unable to locate one genuinely scientific article or work which supported Section 4(a). In order to comply with the mandate of the Board she used such materials as an article from *Reader's Digest* about "atomic clocks" which inferentially suggested that the earth was less than 4½ billion years old. She was unable to locate any substantive teaching material for some parts of Section 4 such as the worldwide flood. The curriculum guide which she prepared cannot be taught and has no educational value as science. The defendants did not produce any text or writing in response to this evidence which they claimed was usable in the public school classroom.

The conclusion that creation science has no scientific merit or educational value as science has legal significance in light of the Court's previous conclusion that creation science has, as one major effect, the advancement of religion. The second part of the three-pronged test for establishment reaches only those statutes having as their *primary* effect the advancement of religion. Secondary effects which advance religion are not constitutionally fatal. Since creation science is not science, the conclusion is inescapable that the *only* real effect of Act 590 is the advancement of religion. The Act therefore fails both the first and second portions of the test in *Lemon v. Kurtzman,* 403 U.S. 602 (1971).

IV.(E)

Act 590 mandates "balanced treatment" for creation science and evolution science. The Act prohibits instruction in any religious doctrine or references to religious writings. The Act is self-contradictory and compliance is impossible unless the public schools elect to forego significant portions of subjects such as biology, world history, geology, zoology, botany, psychology, anthropology, sociology, philosophy, physics and chemistry. Presently, the concepts of evolutionary theory as described in 4(b) permeate the public school textbooks. There is no way teachers can teach the Genesis account of creation in a secular manner.

The State Department of Education, through its textbook selection committee, school boards and school administrators will be required to constantly monitor materials to avoid using religious references. The school boards, administrators and teachers face an impossible task. How is the teacher to respond to questions about a creation suddenly and out of nothing? How will a teacher explain the occurrence of a worldwide flood? How will a teacher explain the concept of a relatively recent age of the earth? The answer is obvious because the only source of this information is ultimately contained in the Book of Genesis.

References to the pervasive nature of religious concepts in creation science texts amply demonstrate why State entanglement with religion is inevitable under Act 590. Involvement of the State in screening texts for impermissible religious references will require State officials to make delicate religious judgments. The need to monitor classroom discussion in order to uphold the Act's prohibition against religious instruction will necessarily involve administrators in questions concerning religion. These continuing involvements of State officials in questions and issues of religion create an excessive and prohibited entanglement with religion.

V.

These conclusions are dispositive of the case and there is no need to reach legal conclusions with respect to the remaining issues. The plaintiffs raised two other issues questioning the constitutionality of the Act and, insofar as the factual findings relevant to these issues are not covered in the preceding discussion, the Court will address these issues. Additionally, the defendants raised two other issues which warrant discussion.

V.(A)

First, plaintiff teachers argue the Act is unconstitutionally vague to the extent that they cannot comply with its mandate of "balanced" treatment without jeopardizing their employment. The argument centers around the lack of a precise definition in the Act for the word "balanced." Several witnesses expressed opinions that the word has such meanings as equal time, equal weight, or equal legitimacy. Although the Act could have been more explicit, "balanced" is a word subject to ordinary understanding. The proof is not convincing that a teacher using a reasonably acceptable understanding of the word and making a good faith effort to comply with the Act will be in jeopardy of termination. Other portions of the Act are arguably vague, such as the "relatively recent" inception of the earth and life. The evidence establishes, however, that relatively recent means from 6,000 to 20,000 years, as commonly understood in creation science literature. The meaning of this phrase, like Section 4(a) generally, is, for purposes of the Establishment Clause, all too clear.

V.(B)

The plaintiffs' other argument revolves around the alleged infringement by the defendants upon the academic freedom of teachers and students. It is contended this unprecedented intrusion in the curriculum by the State prohibits teachers from teaching what they believe should be taught or requires them to teach that which they do not believe is proper. The evidence reflects that traditionally the State Department of Education, local school boards and administration officials exercise little, if any, influence upon the subject matter taught by classroom teachers. Teachers have been given freedom to teach and emphasize those portions of subjects the individual teacher considered important. The limits to this discretion have generally been derived from the approval of textbooks by the State Department and preparation of curriculum guides by the school districts.

Several witnesses testified that academic freedom for the teacher means, in substance, that the individual teacher should be permitted unlimited discretion subject only to the bounds of professional ethics. The Court is not prepared to adopt such a broad view of academic freedom in the public schools.

In any event, if Act 590 is implemented, many teachers will be required to teach material in support of creation science which they do not consider academically sound. Many teachers will simply forego teaching subjects which might trigger the "balanced treatment" aspects of Act 590 even though they think the subjects are important to a proper presentation of a course.

Implementation of Act 590 will have serious and untoward consequences for students, particularly those planning to attend college. Evolution is the cornerstone of modern biology, and many courses in public schools contain subject matter relating to such varied topics as the age of the earth, geology and relationships among living things. Any student who is deprived of instruction as to the prevailing scientific thought on these topics will be denied a significant part of science education. Such a deprivation through the high school level would undoubtedly have an impact upon the quality of education in the State's colleges and universities, especially including the pre-professional and professional programs in the health sciences.

V.(C)

The defendants argue in their brief that evolution is, in effect, a religion, and that by teaching a religion which is contrary to some students' religious views, the State is infringing upon the student's free exercise rights under the First Amendment. Mr. Ellwanger's legislative findings, which were adopted as a finding of fact by the Arkansas Legislature in Act 590, provides:

> Evolution-science is contrary to the religious convictions or moral values or philosophical beliefs of many students and parents, including individuals of many different religious faiths and with diverse moral and philosophical beliefs. Act 590, §7(d).

The defendants argue that the teaching of evolution alone presents both a free exercise problem and an establishment problem which can only be redressed by giving balanced treatment to creation science, which is admittedly consistent with some religious beliefs. This argument appears to have its genesis in a student note written by Mr. Wendell Bird, "Freedom of Religion and Science Instruction in Public Schools," 87 Yale L.J. 515 (1978). The argument has no legal merit.

If creation science is, in fact, science and not religion, as the defendants claim, it is difficult to see how the teaching of such a science could "neutralize" the religious nature of evolution.

Assuming for the purposes of argument, however, that evolution is a religion or religious tenet, the remedy is to stop the teaching of evolution; not establish another religion in opposition to it. Yet it is clearly established in the case law, and perhaps also in common sense, that evolution is not a religion and that teaching evolution does not violate the Establishment Clause.

V.(D)

The defendants presented Dr. Larry Parker, a specialist in devising curricula for public schools. He testified that the public school's curriculum should reflect the subjects the public wants taught in schools. The witness said that polls indicated a significant majority of the American public thought creation science should be taught if evolution was taught. The point of this testimony was never placed in a legal context. No doubt a sizeable majority of Americans believe in the concept of a Creator or, at least, are not opposed to the concept and see nothing wrong with teaching school children about the idea.

The application and content of First Amendment principles are not determined by public opinion polls or by a majority vote. Whether the proponents of Act 590 constitute the majority or the minority is quite irrelevant under a constitutional system of government. No group, no matter how large or small, may use the organs of government, of which the public schools are the most conspicuous and influential, to foist its religious beliefs on others.

The Court closes this opinion with a thought expressed eloquently by the great Justice Frankfurter:

> We renew our conviction that "we have staked the very existence of our country on the faith that complete separation between the state and religion is best for the state and best for religion." *Everson v. Board of Education,* 330 U.S. at 59. If nowhere else, in the relation between Church and State, "good fences make good neighbors."

An injunction will be entered permanently prohibiting enforcement of Act 590.

It is so ordered this January 5, 1982.

WILLIAM R. OVERTON
United States District Judge

NOTES

FOREWORD

1. *21 Scientists Who Believe in Creation*, 2nd ed. (San Diego: Creation-Life Publishers, 1977), p. 8.

INTRODUCTION

1. Henry M. Morris, *The Remarkable Birth of Planet Earth* (Minneapolis: Dimension Books, Bethany Fellowship, Inc., 1972), pp. 99–101.

2. Henry M. Morris, *The Troubled Waters of Evolution* (San Diego: 1974), p. 75.

3. John Paul II, "The Path of Scientific Discovery," *Origins* 11(15 October 1981):279. I am grateful to Sister Kathleen Dolphin for helping me trace this quotation.

4. On this point and for an account of the Arkansas trial, see Stephen Jay Gould, "Moon, Mann, and Otto," *Natural History* 91(March 1982):4, 6, 8, 10. For the views of a theologian who testified at the trial, see Langdon Gilkey, "Creationism: The Roots of the Conflict," *Christianity and Crisis* 42(26 April 1982):108–15. For reflections on the trial by a spectator who lives in Little Rock, see Gene Lyons, "Repealing the Enlightenment," *Harper's* 264 (April 1982):38–40, 73–78. See also the articles in the special section, "The Creationists," *Science* (December 1981):53–60.

CHAPTER 1

1. William Paley, *Natural Theology; Or, Evidences of the Existence and Attributes of the Deity Collected from the Appearances of Nature*, 12th ed. (London: J. Faulder, 1809), p. 441.

2. Roy Porter, *The Making of Geology: Earth Science in Britain, 1660–1815* (Cambridge: Cambridge University, 1977), p. 158. See also Joe D. Burchfield, *Lord Kelvin and the Age of the Earth* (New York: Science History Publications, 1975), pp. 1–20.

3. William Buckland, *Geology and Mineralogy Considered with Reference to Natural Theology*, 2 vols. (London: W. Pickering, 1836), vol. 1, pp. 18–21.

4. See William Coleman, "Lyell and the 'Reality' of Species," *Isis* 53(1962):333–34; and Walter F. Cannon, "The Bases of Darwin's Achievement: A Revaluation," *Victorian Studies* 5(1961):131.

5. Susan Faye Cannon makes this point in *Science in Culture: The Early Victorian Period* (New York: Science History Publications, 1978), pp. 88–90.

6. The phrase was coined by John Herschel, probably the most respected scientist of the early Victorian period. See Walter F. Cannon, "The Problem of Miracles in the 1830s," *Victorian Studies* 4(1960):22–23.

7. Nora Barlow, ed., *The Autobiography of Charles Darwin, 1809–1882* (New York: W. W. Norton, 1969), p. 120. On Darwin's changing ideas from 1835 to 1838, see David Kohn, "Theories to Work By: Rejected Theories, Reproduction, and Darwin's Path to Natural Selection," *Studies in the History of Biology* 4(1980):67–170.

8. See Maurice Mandelbaum, "Darwin's Religious Views," *Journal for the History of Ideas* 19(1958):363–78.

9. Kelvin, "On Mechanical Antecedents of Motion, Heat, and Light," in Kelvin, *Mathematical and Physical Papers*, 6 vols. (Cambridge: Cambridge University, 1882–1911), vol. 2, p. 34.

10. Much of this section is drawn from the discussion of thermodynamics in D. B. Wilson, "Concepts of Physical Nature: From John Herschel to Karl Pearson," in *Nature and the Victorian Imagination*, U. C. Knoepflmacher and G. B. Tennyson, eds. (Berkeley: University of California, 1977), pp. 201–15.

11. See ibid.

12. Kelvin, "On a Universal Tendency in Nature to the Dissipation of Mechanical Energy," *Mathematical and Physical Papers*, vol. 1, pp. 511–14.

13. Kelvin, "The Sorting Demon of Maxwell," *Popular Lectures and Addresses*, 2nd ed. (London: Macmillan, 1891), vol. 1, pp. 144–45.

14. Ibid., p. 145.

15. Ibid., p. 146.

16. Kelvin, "On the Age of the Sun's Heat," *Popular Lectures and Addresses*, vol. 1, p. 357.

17. Kelvin, "The 'Doctrine of Uniformity' in Geology Briefly Refuted," *Proceedings of the Royal Society of Edinburgh* 5(1865):512–13.

18. Benjamin Jowett, "On the Interpretation of Scripture," in Frederick Temple et al., *Essays and Reviews* (London: John W. Parker, 1860), p. 374.

19. Ibid., p. 375.

20. Ibid., p. 376.

21. C. W. Goodwin, "Mosaic Cosmogony," *Essays and Reviews*, p. 217.

22. Ibid., p. 220.

23. Ibid., p. 252.

24. Jowett, "On the Interpretation of Scripture," p. 349.

25. Goodwin, "Mosaic Cosmogony," pp. 214–15.

26. Baden Powell, "On the Study of the Evidences of Christianity," in *Essays and Reviews*, pp. 129, 139.

27. T. H. Huxley, "Agnosticism," *Science and the Christian Tradition* (London: Macmillan, 1894), pp. 218–19.

28. C. L. Drawbridge, ed., *The Religion of Scientists, Being Recent Opinions Expressed by Two Hundred Fellows of the Royal Society on the Subject of Religion and Theology* (New York: Macmillan, 1932), p. 108.

29. Ibid., p. 72.

30. Ibid., p. 78.

31. Ibid., p. 87.

CHAPTER 2

1. I would like to thank Dean R. Fowler for allowing me to read his manuscript, "The Creationist Movement," before publication in the journal *American Biology Teacher*. It helped me trace the relationships of various court decisions and creationists' organizations.

2. H. M. Morris, "Proposals for Science Framework Guidelines," *Creation Research Society Quarterly* 8(September 1971):147–50.

3. H. M. Morris, *The Remarkable Birth of Planet Earth* (Minneapolis: Dimension Books, Bethany Fellowship, Inc., 1972), p. 16.

4. W. T. Stace, *Religion and the Modern Mind* (Philadelphia: Lippincott, 1952), p. 89.

5. Ibid., p. 90.

6. H. M. Morris, *The Troubled Waters of Evolution* (San Diego: Creation-Life Publishers, 1974), p. 184.

7. Jehovah's Witnesses, *Did Man Get Here by Evolution or Creation?* (New York: Watch Tower and Bible Tract Society, 1967).

8. D. L. Willis, ed., *Origins and Change: Selected Readings from the Journal of the American Scientific Affiliation* (Elgin, Illinois: American Scientific Affiliation, 1978), p. 77.

CHAPTER 3

1. S. J. Gould, "Evolution as Fact and Theory," *Discovery* (May 1981):34.

2. Ibid., pp. 34–37.

3. D. Gish, "Evolution as Fact and Theory," *Discovery* (July 1981): 6.

4. S. Sternberg, "Two Operations in Character Recognition: Some Evidence from Reaction-Time Measurements," *Perception and Psychophysics* 2(1967):45–53.

5. More precisely, the mean reaction time and the size of the positive set are linearly correlated.

6. Theories are tested by checking the predicted observational consequences of the theory. Typically, the predicted consequences do not follow *merely* from the theory being tested. The theory must be supplemented with auxiliary assumptions, such as assumptions about background conditions or the apparatus used in the experiment. If, for example, one's theory predicts that a microorganism will increase in size when placed in a certain solution, the test of this view will involve the auxiliary assumption that microscopes faithfully reveal at least the relative size of certain microorganisms. If the predicted consequences fail to occur, it is logically possible that the auxiliary assumptions are false and not the theory proper.

7. Being certain is a logically stronger notion than being correct. If the data are correct, they are true. If the data were certain, they would not merely be true. It would be *impossible* for the data to be false.

8. Typically, phenomena are explained by being shown to be an instance of causal regularities or laws of nature. For the details of scientific explanation and theories, see Part II of Klemke, Hollinger, and Kline and also see Suppe.

9. Gish, "Evolution as Fact and Theory," p. 6.

10. I do not wish to deny that there are interesting differences between the regularities of evolutionary theory and other theories. But those differences do not seem to be at issue in the creationsts' criticism. See D. Hull, *Philosophy of Biological Science* (Englewood Cliffs, N.J.: Prentice-Hall, 1974), pp. 70–100.

11. Whenever any ray of light is incident at the surface that separates two media, it is bent in such a way that the ratio of the sine of the angle of incidence to the sine of the angle of refraction is always a constant quantity for those two media.

12. Very roughly, the product of the pressure and volume is proportional to the temperature.

13. For example, any characteristic will become more prevalent in the population if the individual possessing it produces a larger progeny that survives to adulthood than individuals not having the trait. See Chapter 7 of this volume.

14. See, for example, Gish, "Evolution as Fact and Theory," p. 6.

15. K. Popper, "Science: Conjectures and Refutations," reprinted in *Introductory Readings in the Philosophy of Science*, E. Klemke, R. Hollinger, and A. D. Kline, eds. (Buffalo, N.Y.: Prometheus Books, 1980), pp. 19–34.

16. K. Popper, "Autobiography of Karl Popper," in *The Philosophy of Karl Popper, Book I*, P. Schilpp, ed., (La Salle, Ill.: Open Court, 1974), p. 134.

17. There are serious problems with falsificationism. See P. Thagard, "Why Astrology Is a Pseudoscience," and P. Feyerabend, "How to Defend Society against Science," both in *Introductory Readings in the Philosophy of Science*, E. Klemke, R. Hollinger, and A. D. Kline, eds. (Buffalo, N.Y.: Prometheus Books, 1980).

18. K. Popper, "Evolution," *New Scientist* (August 1980): 611.

CHAPTER 4

1. Henry M. Morris, *The Remarkable Birth of Planet Earth* (Minneapolis: Dimension Books, 1972), p. iv.

2. Ibid., p. v.

3. Ibid., p. 57.

4. Ibid., p. 58.

5. Ibid., p. 59.

6. Ibid., p. 61.

7. Ibid., p. 62.

8. Ibid., pp. 66–67.

9. Ibid., p. 62.

10. C. Wetherill, "The Formation of the Earth from Planetesimals," *Scientific American*, 244 (June 1981):162–74.

11. M. Schwarzschild, *Structure and Evolution of the Stars* (Princeton: Princeton University Press, 1958), p. 1.

CHAPTER 5

1. T. C. Chamberlin, "The Method of Multiple Working Hypotheses," *Journal of Geology* 5(1897):837–48.

2. A. Holmes, *The Age of the Earth* (London and New York: Harper, 1913).

3. C. T. Harper, ed., *Geochronology: Radiometric Dating of Rocks and Minerals* (Stroudsburg, Pa.: Dowden, Hutchinson, and Ross, 1973).

4. J. C. Whitcomb, Jr., and H. M. Morris, *The Genesis Flood,* 23rd printing (Phillipsburg, N.J.: Presbyterian and Reformed Publishing Co., 1979), pp. 213–14.

5. H. M. Morris, *The Beginning of the World* (Denver: Accent Books, 1977), pp. 13–14. It may appear to some readers that we have quoted excessively from the writings of Henry Morris. We wish to justify our utilization of his publications by noting the following points. (1) He is the director of the Institute for Creation Research, and, presumably, his writings accurately portray the position of that institution. ICR publications are used widely within the creationist movement and some have been proposed for use in the public school systems. (2) He has written a considerable volume of material on the geologic aspects of "scientific" creationism, perhaps more than any other modern worker. (3) He is regarded by the present authors as one of the most prolific and influential of the scientific creationists. His writings, the work of other staff members of ICR, and publications of the Creation Research Society form the primary basis for the creationists' interpretation of earth history presented in this chapter.

6. H. M. Morris, "The Tenets of Creationism," *Impact,* no. 85 (San Diego: Institute for Creation Research, 1980).

7. Ibid.

8. Ibid.

9. H. M. Morris, ed., *Scientific Creationism,* gen. ed. (San Diego: Creation-Life Publishers, 1974), p. 149.

10. H. M. Morris, *The Remarkable Birth of Planet Earth* (Minneapolis: Dimension Books, Bethany Fellowship, Inc., 1972), pp. 90–91.

11. Morris, ed., *Scientific Creationism,* pp. 92–94.

12. Morris, *Beginning of the World,* p. 155.

13. H. M. Morris, *The Troubled Waters of Evolution* (San Diego: Creation-Life Publishers, 1975), p. 93.

14. Morris, *Beginning of the World,* p. 110.

15. Morris, ed., *Scientific Creationism,* pp. 104–7.

16. Ibid., pp. 108–10.

17. Ibid., p. 101.

18. Ibid., p. 110.

19. Ibid., p. 117.

20. Morris, *Troubled Waters of Evolution,* p. 93.

21. Morris, ed., *Scientific Creationism,* p. 97.

22. Morris, *Beginning of the World,* p. 110.

23. R. L. Wysong, *The Creation-Evolution Controversy* (Midland, Mich.: Inquiry Press, 1976), p. 361.

24. Morris, *Remarkable Birth of Planet Earth,* p. 77.

25. D. T. Gish, *Evolution? The Fossils Say No!* (San Diego: Creation-Life Publishers, 1973), p. 39.

26. Wysong, *Creation-Evolution Controversy,* p. 352.

27. Morris, ed., *Scientific Creationism,* p. 96.

28. Ibid., p. 136.

29. Morris, *Beginning of the World,* p. 111.

30. Morris, ed., *Scientific Creationism,* p. 120.

31. Morris, *Troubled Waters of Evolution,* p. 96.

32. Morris, *Remarkable Birth of Planet Earth,* p. 28.

33. Morris, ed., *Scientific Creationism,* p. 119.

34. Morris, *Beginning of the World,* p. 112.

35. Morris, *Remarkable Birth of Planet Earth,* p. 89.

36. H. S. Slusher, *Critique of Radiometric Dating,* I.C.R. Technical Monograph 2 (San Diego: Creation-Life Publishers, 1973), p. 43.

37. Ibid., p. 3.
38. J. C. Whitcomb, Jr., *The Early Earth* (Grand Rapids, Mich.: Baker Book House, 1972), p. 33.
39. Morris, *Beginning of the World,* p. 157.
40. Morris, *Remarkable Birth of Planet Earth,* 29.
41. Wysong, *Creation-Evolution Controversy,* pp. 388–92.
42. Morris, *Remarkable Birth of Planet Earth,* p. 28.
43. Ibid., p. 30.
44. Morris, *Beginning of the World,* p. 156.
45. Morris, ed., *Scientific Creationism,* pp. 117–18.
46. Whitcomb and Morris, *The Genesis Flood,* note 4, p. 213.
47. Morris, ed., *Scientific Creationism,* p. 94.
48. Morris, *Remarkable Birth of Planet Earth,* p. 91.
49. Morris, ed., *Scientific Creationism,* p. 117.
50. Ibid., p. 120.
51. Morris, *Troubled Waters of Evolution,* p. 96.
52. R. H. Dott, Jr., and R. L. Batten, *Evolution of the Earth,* 2nd ed. (New York: McGraw-Hill, 1976), p. 38.
53. W. Buckland, *Geology and Mineralogy Considered with Reference to Natural Theology* (Philadelphia: Carey, Lea, and Blanchard, 1837), p. 441.
54. S. J. Gould, "Is Uniformitarianism Necessary?" *American Journal of Science* 263 (1965):223–28.
55. Whitcomb and Morris, *The Genesis Flood,* pp. 123–24.
56. Ibid.
57. Ibid., pp. xxvi–xxvii.
58. Morris, *Beginning of the World,* p. 108.
59. Morris, ed., *Scientific Creationism,* p. 103.
60. Whitcomb and Morris, *The Genesis Flood,* p. 145.
61. See Whitcomb, *The Early Earth,* note 38, p. 37.
62. Ibid., p. 91.
63. Morris, *Beginning of the World,* pp. 149, 151.
64. Ibid.
65. H. M. Morris, *Evolution and the Modern Christian* (Phillipsburg, N.J.: Presbyterian and Reformed Publishing Co., 1967), p. 40.
66. Whitcomb and Morris, *The Genesis Flood,* p. 165; Wysong, *Creation-Evolution Controversy,* note 23, p. 368.
67. For example, J. C. Crowell, "Gondwana Glaciation, Cyclothems, Continental Positions, and Climate Change," *American Journal of Science* 278(1978):1345–73.
68. Whitcomb and Morris, *The Genesis Flood,* p. 248.
69. Morris, ed., *Scientific Creationism,* p. 104.
70. Ibid., p. 105.
71. J. L. Wilson, *Carbonate Facies in Geologic History* (New York: Springer-Verlag, 1975), p. 18.
72. Whitcomb and Morris, *The Genesis Flood,* p. 409.
73. D. B. D'Armond, "Thornton Quarry Deposits: A Fossil Coral Reef or a Catastrophic Flood Deposit? A Preliminary Study," *Creation Research Society Quarterly* 17(1980):103.
74. H. L. Penman, "The Water Cycle," *Scientific American,* 223(1970):98–108; P. Cloud and A. Gibor, "The Oxygen Cycle," *Scientific American* 223(1970):111–23; B. Bolin, "The Carbon Cycle," *Scientific American* 223(1970):124–32.
75. R. G. C. Bathurst, *Carbonate Sediments and Their Diagenesis,* 2nd ed. (Amsterdam: Elsevier Publications, 1975), p. 231.
76. H. Blatt, G. Middleton, and R. Murray, *Origin of Sedimentary Rocks* (Englewood Cliffs, N.J.: Prentice-Hall, 1972).
77. Morris, ed., *Scientific Creationism,* p. 105.
78. Whitcomb and Morris, *The Genesis Flood,* p. 412.
79. Morris, ed., *Scientific Creationism,* p. 107.
80. Ibid., p. 109.

81. Ibid., pp. 109–10.
82. Bolin, "The Carbon Cycle," note 74.
83. Gish, *Evolution? The Fossils Say No!* note 25, p. 39.
84. Dott and Batten, *Evolution of the Earth,* note 52, p. 47.
85. See ibid., pp. 21–24.
86. Whitcomb and Morris, *The Genesis Flood,* p. 168.
87. K. W. Stockman, R. N. Ginsburg, and E. A. Shinn, "Production of Lime Mud by Algae in South Florida," *Journal of Sedimentary Petrology* 37(1967):633–48.
88. P. Enos and R. D. Perkins, "Evolution of Florida Bay from Island Stratigraphy," *Geological Society of America: Bulletin* 90(1979):59–83, Fig. 4.
89. Morris, *Troubled Waters of Evolution,* note 13, p. 93.
90. W. B. N. Berry, *Growth of a Prehistoric Time Scale, Based on Organic Evolution* (San Francisco: Freeman, 1968), pp. 116–17.
91. J. E. Repetski, "A Fish from the Upper Cambrian of North America," *Science,* 200(1978):529–31.
92. Morris, ed., *Scientific Creationism,* p. 117.
93. Whitcomb and Morris, *The Genesis Flood,* p. 327.
94. P. E. Cloud, Jr., "Pre-Metazoan Evolution and the Origins of the Metazoa," in *Evolution and Environment,* E. T. Drake, ed. (New Haven and London: Yale University Press, 1968), pp. 1–72.
95. Whitcomb and Morris, *The Genesis Flood,* pp. 66–67.
96. N. Eldredge and S. J. Gould, "Punctuated Equilibria: An Alternative to Phyletic Gradualism," in *Models in Paleobiology,* T. J. M. Schopf, ed. (San Francisco: Freeman, Cooper, 1972), pp. 82–115.
97. B. A. Malmgren and J. P. Kennett, "Phyletic Gradualism in a late Cenozoic Foraminiferal Lineage; DSDP Site 284, Southwest Pacific," *Paleobiology* 7(1981):230–40.
98. D. J. Jones, "Displacement of Microfossils," *Journal of Sedimentary Petrology* 28(1958):453–67.
99. For example, G. Klapper, "Upper Devonian and Lower Mississippian Conodont Zones in Montana, Wyoming, and South Dakota," *University of Kansas Paleontological Contributions* paper 3 (1966):12.
100. For example, G. Seddon, "Pre-Chappel Conodonts of the Llano Region, Texas," *Texas University Bureau of Economic Geology: Report of Investigations* 68(1970):76–78.
101. Whitcomb and Morris, *The Genesis Flood,* pp. 206–7.
102. R. Buick, J. S. R. Dunlop, and D. I. Groves, "Stromatolite Recognition in Ancient Rocks: An Appraisal of Irregularly Laminated Structures in an Early Archaean Chert-Barite Unit from North Pole, Western Australia," *Alcheringa* 5(1981):161–81; D. I. Groves, J. S. R. Dunlop, and R. Buick, "An Early Habitat of Life," *Scientific American* (October 1981): 64–73.
103. Cloud, "Pre-Metazoan Evolution," note 94.
104. A. H. Knoll and G. Vidal, "Early Plankton: Radiations and Extinctions in the Late Precambrian and Early Cambrian," *Geological Society of America: Abstracts with Programs* 13(1981):489.
105. M. F. Glaessner and M. Wade, "The Late Precambrian Fossils from Ediacara, South Australia," *Palaeontology* 9(1966):599–628; J. W. Cowie and A. Yu. Razanov, "I.U.G.S. Precambrian/Cambrian Boundary Working Group in Siberia, 1973," *Geological Magazine* 111 (1974):237–52.
106. Gish, *Evolution? The Fossils Say No!* note 25, p. 45; Glaessner and Wade, "The Late Precambrian Fossils."
107. For example, Whitcomb and Morris, *The Genesis Flood,* pp. 172–73.
108. B. Neufeld, "Dinosaur Tracks and Giant Men," *Origins* 2(1975):64–76.
109. J. D. Morris, *Tracking Those Incredible Dinosaurs . . . and the People Who Knew Them* (San Diego: Creation-Life Publishers, 1980).
110. Neufeld, "Dinosaur Tracks and Giant Men," p. 75.
111. C. J. Jolly and F. Plog, *Physical Anthropology and Archeology* (New York: Alfred A. Knopf, 1976).
112. Slusher, *Critique of Radiometric Dating,* note 36.
113. T. W. Rybka, "Consequences of Time Dependent Nuclear Decay Indices on Half Lives,"

Impact, no. 106 (San Diego: Institute for Creation Research, 1982).

114. S. G. Brush, "Finding the Age of the Earth: By Physics or by Faith?" *Journal of Geological Education* 30(1982):35.

115. R. L. Fleischer and P. B. Price, "Techniques for Geological Dating of Materials by Chemical Etching of Fission Fragment Tracks," *Geochimica et Cosmochimica Acta* 28(1964): 1705–14.

116. Morris, ed., *Scientific Creationism,* p. 149.

117. Ibid.

118. T. G. Barnes, *Origin and Destiny of the Earth's Magnetic Field* (San Diego: Institute for Creation Research, 1973).

119. Brush, "Finding the Age of the Earth," p. 36.

120. E. Irving, *Paleomagnetism and Its Applications to Geological and Geophysical Problems* (New York: Wiley, 1964).

CHAPTER 7

1. H. M. Morris, *Scientific Creationism* (San Diego: Creation-Life Publishers, 1974), pp. 216–18.

2. J. C. Whitcomb, Jr., and H. M. Morris, *The Genesis Flood* (Phillipsburg, N.J.: Presbyterian and Reformed Publishing Co., 1961), pp. 265–66.

3. H. M. Morris, *Scientific Creationism,* p. 122.

4. R. E. Sloan, "The Association of 'Human' and Fossil Footprints," *Journal of the Minnesota Science Teachers Association* 1(1979):45–46.

5. Whitcomb and Morris, *The Genesis Flood,* pp. 270–88.

6. G. L. Stebbins and F. J. Ayala, "Is a New Evolutionary Synthesis Necessary?" *Science* 213(1981):967–71.

7. J. N. Moore, "Evolution, Creation, and the Scientific Method," *American Biology Teacher* 35(1973):23–26.

8. G. L. Stebbins, *Processes of Organic Evolution* (Englewood Cliffs, N.J.: Prentice-Hall, 1977), pp. 30–31.

9. W. E. Lammerts, *Why Not Creation?* (Grand Rapids, Mich.: Baker Book House, 1981), pp. 243–67.

10. H. M. Morris, *Scientific Creationism,* p. 79.

11. T. Dobzhansky, "Nothing in Biology Makes Sense Except in the Light of Evolution," *American Biology Teacher* 35(1973):125–29.

CHAPTER 8

1. Duane T. Gish, *Evolution? The Fossils Say No!* (San Diego: Creation-Life Publishers, 1973).

2. Ibid., p. 107.

3. Ibid., p. 108.

4. Ibid., p. 111.

5. In conventional biological nomenclature, a species is labeled with its generic as well as its specific Latin names. The genus label comes first, has an initial capital letter and is followed by the species name. Both are italicized. When a trinomial label is used, the third name, also italicized, refers to a subspecies. Obsolete names are shown within quotation marks, and their replacements follow immediately in parentheses. Thus *"Plesianthropus"* (= *Australopithecus*) means that a group of fossils that used to be considered a separate genus, *"Plesianthropus,"* is now subsumed in the genus *Australopithecus.* The main reason I have included obsolete names in this chapter is that they appear in the creationist literature cited among my references.

6. Glynn E. Daniel, "The Idea of Man's Antiquity," in *Avenues to Antiquity,* Brian M. Fagan, ed. (San Francisco: Freeman, 1976), pp. 9–13. While the original calculations by Archbishop Ussher of Armagh placed the date of creation very precisely at 4004 B.C., modern creationists apparently allow a greater margin for error, since they refer to an interval between six and ten thousand years ago.

7. The average brain size of modern humans is approximately 1,400 cubic centimeters, with

a range from about 1,000 to 2,000 cubic centimeters, while ape brains average around 500 cubic centimeters, with a range from about 280 to 750 cubic centimeters. Since brain size is to some extent correlated with body size in humans and their near relatives, a more meaningful comparison would involve the ratio of brain weight to body weight. Here, the contrasts are striking — 1:47 for modern humans and 1:135 for chimps, who have the highest values for relative brain weight among the apes.

8. The Miocene epoch spans the period from about twenty-five to five million years ago, and the Pliocene runs from five to two million years before the present.

9. The Pleistocene epoch covers the period from about two million to ten thousand years ago and includes the great cycles of continental glaciation known as the Ice Age.

10. J. S. Weiner, *The Piltdown Forgery* (London: Oxford University Press, 1955).

11. I have placed the term "race" within quotation marks because its applicability to the human species is seriously questioned by many physical anthropologists. The interested reader may wish to pursue this topic in various of the books listed at the end of the chapter.

12. Potassium-argon dating is the technique of measuring the amount of radioactive disintegration recorded in volcanic rocks since the time of their eruption. It has been widely and convincingly applied to deposits at many of the African sites.

CHAPTER 9

1. A priori here means that the probabilities are calculated on the basis of the number of alternative ways a system can achieve a given state as determined, for example, by variables such as pressure, temperature, and volume.

2. A mole of oxygen contains 6×10^{23} O_2 molecules.

CHAPTER 10

1. James E. Strickling, "Creation, Evolution, and Objectivity," *Creation Research Quarterly* 16(1979):98–100.

2. The following translations are based on E. A. Speiser, ed., *Genesis* (Garden City, N.Y.: Doubleday, 1964).

3. Ibid.

4. But see the flood story, Genesis 7:11, in which the "windows of the heavens were opened" in order to help flood the earth.

5. T. H. Gaster, "Cosmogony," in *Interpreter's Dictionary of the Bible* (Nashville: Abingdon Press, 1962), Fig. 50.

6. Mircea Eliade, *Gods, Goddesses, and Myths of Creation* (New York: Harper & Row, 1967), pp. 84–85, 97–98.

7. Walter Harrelson, *From Fertility Cult to Worship* (Garden City, N.Y.: Doubleday, 1969), p. 76.

8. See John S. Hawley, "Krishna's Cosmic Victories," *Journal of the American Academy of Religion* 47(1979):201–21. It may also be noted in this connection that all other references to creation in the Bible are also of a religious character, however much they may differ from Genesis in details. Examples would be some of the Psalms (8, 19, 104, and so on), prophets (for example, Isaiah 41:17–20; 42:14–17; 43:1–7, 16–21; 44:2,21, 24; 45:9–13; 48:12–13; 51:8–11), wisdom (Job 26, 38), and the New Testament (for example, Ephesians 1:9–11; Colossians 1:15–17; John 1:1–18).

CHAPTER 11

1. Winthrop S. Hudson, *Religion in America* (New York: Charles Scribner's Sons, 1973), p. 265.

2. As quoted in ibid., p. 267.

3. Sydney E. Ahlstrom, *A Religious History of the American People* (New Haven, Conn.: Yale University Press, 1972), p. 909.

4. As quoted in ibid., p. 909.

5. James Barr, *Fundamentalism* (Philadelphia: Westminster Press, 1978), p. 40.

6. Harold Lindsell, *The Battle for the Bible* (Grand Rapids, Mich.: Zondervan, 1976), p. 18.

7. Barr, *Fundamentalism*, p. 40.

8. Augustus Strong, *Systematic Theology* (New York: A. C. Armstrong and Son, 1896), p. 193.

9. Ibid., p. 194.

10. Bernard Ramm, *The Christian View of Science and Scripture* (Grand Rapids, Mich.: William B. Eerdmans, 1954), p. 25.

11. Ibid., p. 152.

12. As quoted in ibid., p. 200.

13. Emil Brunner, *The Christian Doctrine of God* (Philadelphia: Westminster Press, 1950), p. 176.

14. John B. Cobb, Jr., *God and the World* (Philadelphia: Westminster Press, 1969), p. 65.

15. Robert B. Mellert, *What Is Process Theology?* (New York: Paulist Press, 1975), p. 55.

16. Pierre Teilhard de Chardin, *The Phenomenon of Man* (New York: Harper & Row, 1959), p. 219.

17. Ibid., p. 39.

18. Ibid., p. 86.

CHAPTER 12

1. Donald E. Miller, *The Case for Liberal Christianity* (New York: Harper & Row, 1981), p. 85.

2. Ibid., p. 77.

CHAPTER 13

1. 89 S. Ct. 272 (1968).

2. Scopes v. Tennessee, 154 Tenn. 105, 289 S.W. 363 (1927).

3. Leo Pfeffer, "Religion, Education and the Constitution," *Lawyer's Guild Review* 8(1948):387.

4. Peter G. Mode, *Sourcebook and Bibliographical Guide for American Church History* (Menasha, Wis.: Banta, 1921), pp. 78–79.

5. R. F. Butts, *American Tradition in Religion and Education* (Boston: Beacon Press, 1950), p. 113. See also D. G. Boorstin, *The Americans* (New York: Random House, 1958), pp. 117–19; and C. H. Dodd, *The Authority of the Bible* (New York: Harper Torch Book, 1958), pp. 173–76.

6. Merle Curti, *The Growth of American Thought* (New York: Harper, 1951), p. 53.

7. See M.S. Bates, *Religious Liberty: An Inquiry* (New York: Harper, 1945).

8. E. P. Cubberley, *Public Education in the United States* (Boston: Houghton Mifflin, 1934).

9. Leo Pfeffer, *Church, State and Freedom* (Boston: Beacon Press, 1953).

10. Nathan Schachner, *Church, State and Education*, reprinted from *The American Jewish Yearbook (1947–48),* vol. 49, (New York: American Jewish Committee, 1947), p. 11.

11. Ibid.

12. Sherman Smith, *The Relation of the State to Religious Education* (New York, 1926), p. 71.

13. Cubberley, *Public Education in the United States,* p. 41.

14. Curti, *The Growth of American Thought,* p. 54.

15. M. G. Hunt, "Bible Study in Public Schools," *Peabody Journal of Education* 23 (November 1945):156.

16. Curti, *The Growth of American Thought,* p. 55.

17. C. H. Moehlman, *School and Church* (New York: Harper, 1944), p. 28.

18. Ibid.

19. Curti, *The Growth of American Thought,* pp. 57–58.

20. Moehlman, *School and Church,* p. 27.

21. Charles Beard and Mary Beard, *The Rise of American Civilization* (New York: Macmillan 1933).

22. W. Haller, "The Puritan Background of the First Amendment," in *The Constitution*

Reconsidered, Conyers C. Read, ed. (New York: Columbia University Press, 1938).

23. 330 U.S. 1 (1947) and McCullom v. Board of Education; see also 33 U.S. 203 (1948).

24. 370 U.S. 421 (1962).

25. Abington School District v. Schempp, 374 U.S. 203 (1963).

26. This position is excellently summarized in R. F. Butts, *The American Tradition in Religion and Education* (Boston: Beacon Press, 1949); in Justice Wiley B. Rutledge's dissenting opinion in the Everson case; and in the majority and concurring opinions of the Supreme Court in the Engel and Schempp cases.

27. This viewpoint can be seen in James O'Neill, *Religion and Education under the Constitution* (New York: Harper, 1949), and in a much more restrained fashion in the dissenting opinion of Justice Stewart in the Engel and Schempp cases. Also worthy of note in this connection is E. S. Corwin, *A Constitution of Power in a Secular State* (Charlottesville, Va.: Michie, 1951).

28. See, for example, Paul Wilstach, *The Adams-Jefferson Letters* (Indianapolis: Bobbs-Merrill, 1925); and R. M. Healey, *Jefferson on Religion and Public Education* (New Haven, Conn.: Yale University Press, 1962).

29. I. Brant, *James Madison — The Nationalist, 1780–1787* (Toronto: McCelland & Stewart, 1948).

30. For discussion of this development see C. H. Moehlman, *The Wall of Separation between Church and State* (Boston: Beacon Press, 1951); and Moehlman, *School and Church.*

31. This development is traced in Roy F. Nicholas, *Religion in American Democracy* (Baton Rouge: Louisiana State University Press, 1959); and Charles B. Kinney, Jr., *Church and State, The Struggle for Separation in New Hampshire, 1630–1900* (New York: Teachers College Press, Columbia University, 1955).

32. See R. A. Billington, *The Protestant Crusade, 1800 – 1860* (New York: Macmillan, 1938).

33. See A. P. Stokes, *Church and State in the United States* (New York: Harper and Brothers, 1950), vol. 1, p. 68.

34. Pfeffer, *Church, State and Freedom,* p. 142.

35. Ibid., p. 142.

36. Ibid., p. 392.

37. For discussion of this see H. Beale, *A History of Freedom of Teaching in American Schools* (New York: Octagon, 1941), p. 214.

38. For a thorough discussion of this and the history of the development of public schools, see Donald E. Boles, *The Bible, Religion and the Public Schools* (Ames: Iowa State University Press, 1965).

39. U.S. Stat. at large, 29; 411 (June 11, 1896).

40. *The State and Sectarian Education,* National Educational Association Research Bulletin (Washington, D.C., February 1946), p. 36.

41. See Scopes v. State, 154 Tenn. 105, 289 S.W. 363 (1927).

42. Daniels v. Waters, 515 F.2d 485 (6th Cir. 1975).

43. 45 L.W. 2530 (May 17, 1977).

44. Ibid.

45. Reynolds v. United States, 98 U.S. 145 (1879).

46. See for example: Two Guys from Harrison-Allentown Inc. v. McGinley, 366 U.S. 582 (1961); Gallagher v. Crown Kosher Market, 366 U.S. 617 (1961); and especially Braunfeld v. Brown, 366 U.S. 599 (1961).

47. Torcaso v. Watkins, 367 U.S. 488 (1961).

48. 374 U.S. 398 (1963).

49. 101 S.C. 1425 (1981).

50. 406 U.S. 205 (1972).

51. 374 U.S. 398 (1963).

52. Engel v. Vitale, 37 U.S. 421 (1962).

53. Abington School District v. Schempp, 374 U.S. 203 (1963).

54. See, for example, Lemon v. Kurtzman, 403 U.S. 602 (1971); Hunt v. McNair, 413 U.S. 734 (1972); and Stone v. Graham, 101 S. Ct. 192 (1981).

55. 330 U.S. 1 (1947).

56. See Thomas v. Review Board, 101 S. Ct. at 1435 (1981).

57. McLean v. Arkansas Board of Education, 50 L.W. 2412 (1982).

CHAPTER 14

1. For example, Lawrence Root, "Modern Creationists Seeking Equal Time in the U.S. Classrooms," *Wall Street Journal,* 15 June 1979; "Creationism Evolves," *Scientific American* 240(July 1979):72; *New York Times,* 25 November 1979; *U.S. News and World Report,* 9 June 1980; Jack A. Gerlovich et al., "Creationism in Iowa," *Science* 208(June 1980):1212.

2. Stanley L. Weinberg, "Reactions to Creationism in Iowa," *Creation/Evolution* 2(1980):1; Kansas Association of Biology Teachers, "Creation, Evolution and Public Education: The Proposed Position of the Kansas Association of Biology Teachers," *KABT Newsletter* 21(August 1980):15–18.

3. Sherry Ricchiardi, "Evangelicals Arise on Campus," *Des Moines Sunday Register,* 26 November 1978, section C; Bonnie Wittenberg, "ISU Teacher Denies He Made Antievolution Remark," *Des Moines Register,* 12 April 1979; D. Vance Hawthorne, "Panel Likely to Endorse the Renomination of Koons," *Des Moines Register,* 31 January 1980; Jim Healey, Sherry Ricchiardi, and D. Vance Hawthorne, "ISU Bible Study Group: 'Wonderful' or a 'Cult,' " *Des Moines Register,* 9 March 1980; D. Vance Hawthorne and Jim Healey, "4 Bible Study Members Taken Off ISU News Board," *Des Moines Register,* 13 March 1980; Vyrl Cadwell, "Mike O'Malley and Jeff Nelson, ISU Bible Study Defenders, Say It's Not a Cult," *Des Moines Register,* 23 March 1980; Lisa Theobald, "Student May Appeal Baker Decision to Parks," *Iowa State Daily,* 19 June 1980.

4. Iowa General Assembly, House File 154, by Horace Daggett (Des Moines, February 1977).

5. For example, Dorothy Nelkin, *Science Textbook Controversies and the Politics of Equal Time* (Cambridge: MIT Press, 1977).

6. E. Edward Reutter, Jr., and Robert R. Hamilton, *The Law of Public Education* (Mineola, N.Y.: Foundation Press, 1976), p. 55.

7. Iowa Department of Public Instruction, Curriculum Division, *Creation, Evolution and Public Education: The Position of the Iowa DPI* (Des Moines, 1978).

8. Jack A. Gerlovich, "Creation, Evolution and Public Education," *Iowa Science Teachers Journal,* 15(September 1978):3–6; P. Swami, ed., "Creation, Evolution and Public Education: The Position of the Iowa Department of Public Instruction," *Capsule: Council of State Science Supervisors,* 12(March 1978):1–10; P. Swami, ed., "Iowa's Position on Creation/Evolution Education," *Science Education News* (Summer 1978):2.

9. Hendren v. Campbell, 557-0139, Sup. Ct. 5 (Ind. 1977).

10. "Equal Time for Hokum," *Des Moines Register,* 2 February 1979; "Academic Freedom?" *Des Moines Register,* 29 March 1979.

11. Iowa General Assembly, Senate File 261, by John W. Jensen (Des Moines, February 1979).

12. S. Walters, *Des Moines Register,* 5 April 1979.

13. Iowa General Assembly, Senate File 458, by John W. Jensen (Des Moines, March 1979).

14. State of Iowa, Fiscal Note Worksheet, Request No. 79–269, 1 March 1979.

15. Iowa Academy of Science, *Resolution* (Cedar Falls: University of Northern Iowa, 1979).

16. H. D. Hagen, *Response to Senator B. Van Gilst* (Des Moines: Office of the Iowa Attorney General, August, 28, 1979).

17. Iowa General Assembly, Senate File 458, by Raymond Taylor (Des Moines, February 1980).

18. Iowa General Assembly, Senate File 280, by Raymond Taylor (Des Moines, January 1981).

19. *Position Statement on the Creation/Evolution Controversy from the Iowa Council of Science Supervisors (CS²),* Iowa Department of Public Instruction (Des Moines, 1981).

20. Iowa Academy of Science, *Statement of the Position of the Iowa Academy of Science on the Status of Creationism as a Scientific Explanation of Natural Phenomena* (Cedar Falls: University of Northern Iowa, 1981).

21. National Association of Biology Teachers, *A Compendium of Information on the Theory of Evolution and the Evolution-Creationism Controversy* (Reston, Va., 1977).

22. Robert Marantz Henig, "Evolution Called a 'Religion,' Creationism Defended as 'Science,' " *BioScience* 29(September 1979):513.

23. "Putting Darwin Back in the Dock," *Time,* 16 March 1981, p. 80.

24. Weinberg, "Reactions to Creationism in Iowa."

INDEX

Adam and Eve, 22, 30, 33
Adams, John, 178
Adler, Alfred, 41
Ahlstrom, Sidney E., 149
American Civil Liberties Union, 192, 200
American Institute for Biological Sciences, 20
American Jewish Committee, 187
American Jewish Congress, 187
American Scientific Affiliation, 21, 34, 206
Amish, 183, 189
Archeopteryx, 113
Arkansas Board of Education, 206
Arkansas creationism trial, xxi, 186, 187, 200, 206–22
Arkansas Department of Education, 209, 219, 220
Arkansas Education Association, 187
Arkansas legislature, 208–10, 221
Ayala, Francisco J., 213

Barnes, T. G., 82, 83
Beard, Charles and Mary, 177
Beecher, Henry Ward, 149
Benton, Robert D., 191
Berkeley, William, 173
Bethell, Tom, 207, 209
Bible Science Association, 21, 206
Bible-Science Radio, 21
Biblical scholarship, xix, 12–14, 16, 17, 138–52, 163
Big bang theory, 48, 154
Biological Sciences Curriculum Study, 20
Bird, Wendell, 24, 207, 221
Black, Hugo L., 184
Blackmun, Harry A., 185
Blaine, James G., 179
Bliss, Richard, 24, 206, 217
Blount, W. A., 209
Bolin, B., 73
Boylan, David R., 190
Brant, I., 178
Brongniart, Alexandre, 74
Brunner, Emil, 156
Brush, S. G., 82
Bryan, William Jennings, 150
Buckland, William, 5–6
Burger, Warren, 183, 185

California Board of Education, 21
Calvin, John, 172
Calvinism, 163, 172, 175
Carlucci, Joseph, 208
Carnell, E. J., 67
Carroll, Lewis, 43
Carter, Jimmy, 198
Catholicism, xx, 150, 155, 178–80, 187, 189
Central Baptist College, 217
Christian Heritage College, 21, 23, 25, 206
Citizens against Federal Establishment of Evolutionary Dogma, 21
Citizens for Fairness in Education, 21, 207
Clark, Martin E., 187, 206, 207
Clark, Tom C., 184
Cobb, John B., Jr., 157
Committees of Correspondence, 199–200
Connecticut Constitution, 176
Constitutional Convention, 177
Copernicus, Nicolaus, 85, 159
Creation-Life Publishing Company, 24, 206
Creation Research Society, vii, 24, 60
  belief statement required of members, 22, 138–39, 142, 146, 215
  origin, 21, 206
Creation Science Research Center, 21, 23, 206, 207
Creation stories, xix, xx, xxii, 115, 138–46, 157, 159, 163, 211
Crick, Francis, 89
Cubberley, E. P., 174
Curti, Merle, 173, 175
Cuvier, Georges, 4, 7, 74

Dana, James D., 153, 154
D'Armond, D. B., 71
Darrow, Clarence, 150
Darwin, Charles, xviii, 3, 4, 6, 149
  Age of Darwin, xv, 17
  age of earth, 11, 12, 73
  Essays and Reviews, 13
  Indiana Superior Court and his theory, 181
  natural selection, 8, 107, 109
  origin of life, 87
  Origin of Species, xiv, 10, 19, 73, 152
  religious view, 8–9
  response to critics, 15, 16

235

Darwin, Charles (*continued*)
  suported by T. H. Huxley, 15
  Voyage of H.M.S. *Beagle,* 7
da Vinci, Leonardo, 66
Declaration of Independence, 177
Demons, evil spirits, viii, 15, 47
Descartes, René, 13
Design argument, 4, 6, 9, 12, 14–16, 26, 31, 149
Des Moines Register, 194, 196
Devil, xi, 174, 208
Dobzhansky, Theodosius, 113
Drawbridge, C. L., 16–17

Earth
  age, xvii, 28, 56–57, 87, 221
    atomic clocks, 219
    biblical value, 5, 23, 25, 29, 30, 60, 65, 213, 214, 220
    earth's magnetic field, 64, 82–83
    Kelvin's estimates, 11–12, 16
    radiometric dating, xiv, 58, 63, 81, 82, 216
  history, xiv, xvii, 4–7, 11–12, 55–83
  origin, xvii, 5, 28, 30, 52–54, 207
Eigen, Manfred, 97
Einstein, Albert, 41, 50, 101
Eliade, Mircea, 143
Ellwanger, Paul, 21, 197, 207–9, 214, 221
Epperson, Susan, 20, 170, 181, 210
Evil spirits. *See* Demons

Falsifiability, xvi, 40–42, 213
Family, Life, America under God (FLAG), 208
Fisher, Larry, 216, 217
Fortas, Abe, 170, 209
Fossil record, xiv, 4–6, 32, 42, 56–66, 68–71, 74–76
  evolution, human, xviii, 114–26
  evolution of life, xvii, 7, 16, 25, 27–29, 65, 76–79, 83, 104–7, 111–13
  footprints, human, 80–81, 104
  origin of life, xviii, 87
  Precambrian fossils, 80
  reworked fossils, 77, 80
Fox, S. W., 95
Frankfurter, Felix, 222
Freud, Sigmund, 41

Galileo, 3
Garibaldi, Giuseppe, 179
Geisler, Dr. Norman (witness), 212, 216
Gentry, Robert, 216
Gilkey, Langdon, 212
Gish, Duane, 21, 38–40, 80, 114, 115, 206, 207, 212, 214, 218
Gladden, Washington, 165
Glaessner, M. F., 80
Gore, Charles, 13–14
Gould, Stephen Jay, 38, 213, 215

Grant, U.S., 179
Greater Little Rock Evangelical Fellowship, 209
Guyot, Arnold, 153, 154

Haldane, J. B. S., 87
Harlan, John Marshall, 184
Hays, Robert E., 207
Hodge, Charles, 149
Holsted, James L., 209
Hudson, Winthrop S., 149
Human evolution, xviii, 9, 13, 25, 111, 114–26, 217
Hume, David, 14, 15, 162
Hunt, Carl, 209
Hutton, James, 57, 66
Huxley, T. H., 14–16

Indiana Superior Court, 24, 181
Indiana Textbook Commission, 181
Institute for Creation Research, 21, 60, 190, 206, 207
  origin, 23
  publications, 24, 25, 46, 64, 212, 217
  research, 23
Iowa Academy of Science, 190, 192, 196, 198, 200, 203–4
Iowa Academy of Science Panel on Controversial Issues, 199, 200
Iowa Board of Regents, vii, 190
Iowa Council of Science Supervisors (CS²), 197, 204–5
Iowa Department of Public Instruction, 190–97, 200, 203
Iowa legislature, viii, 189–91, 193, 194, 196–98
Iowa State University, vii–ix, 37, 190, 196
Isely, Duane, ix

Jefferson, Thomas, 178, 181
Jehovah's Witnesses, 34, 183
Jensen, John W., 200–202
Jesus, vii, xi, xv, 14, 15, 22, 34, 142, 162, 164–67, 208, 218
John Paul II, xx
Jones, J. C., 155
Judaism, xix, 24, 43, 115, 180, 186, 189, 208, 218

Kant, Immanuel, 162–63
Keith, Bill, 208
Kelvin, Lord, 3, 10–12, 16
Kennedy, John, 38
Kennett, J. P., 77
Kofahl, Robert E., 207

Lamarck, Jean, 4, 9, 16
Lammerts, Walter, 21
Leakey, Richard, 115, 126

Life
  evolution, xviii, 25–29, 76–77, 80, 83,
    103–13
    Christian views of, xv, xix, xx, 13, 16,
      34–35, 148–59
    fact or theory, xvi, 37–39
    lack of theological implications, xv, 26,
      212
    nineteenth-century theories, xiv, 4,
      7–13, 15–16, 87, 107, 109
    survey of Royal Society of London,
      16–17
    thermodynamics, xix, 31, 127, 134
  origin, xvii–xix, 25, 28, 85–101, 207, 212,
    214–16
    Kelvin's meteoric hypothesis, 11
    thermodynamics, 24, 90, 91, 127,
      132–34
Lindsell, Harold, 150
Lowell, James, 176
Lyell, Charles, 3, 5–7, 9, 66

Madison, James, 177, 178
Magrane, George, 199
Malmgren, B. A., 77
Malthus, Thomas, 7–8
Mann, Horace, 178
Marx, Karl, 41
Maryland Constitution, 183
Massachusetts Constitution, 174
Materialistic determinism, xiii, xv, 34
Maxwell, James Clerk, 10
Mencken, H. L., 172
Mennonites, 189
Mill, John Stuart, 14–16
Miller, Mary Ann, 208
Miller, Stanley, 92
Modernist-fundamentalist controversy, 148–50
Moore, J. N., 24, 111
Moore, John W., 206
Moral Majority, 21
Morris, Henry M., 81, 206
  bible as a source of knowledge, 46, 215
  contrasts evolution and creationism, 25
  creationist institutions, 21, 23, 24
  earth history, 60, 67, 70–72
  Devil, xi
  evil spirits, 47
  fossil record, 32, 63
  morality, 33, 187, 207, 218
  origin of universe, 212
  stellar evolution, xvii, 46–47
Morris, John D., 81
Moses, 13
Muglenberg, Henry Melchior, 173

Nast, Thomas, 179
National Association of Biology Teachers, 20,
  187
National Park Service, 21
National Science Foundation, 20, 21, 216
Neufeld, B., 81

Newton, Isaac, 10, 13, 40, 101
Noachian Flood, xi, xviii, 23, 30, 60, 64–73,
  207, 213, 216
  Arkansas Creationism Act, 211, 214, 219
  Creation Research Society's belief state-
    ment, 22
  fossil record, 32, 58, 68, 70, 71
  nineteenth-century views, 5–6
  "post-Flood man," 115
North Carolina Constitution, 175

Odysseus, 35
Oparin, A. I., 87
Orgel, Leslie, 97
Oswald, Lee Harvey, 38
Overton, William R., xxi, 200, 206, 222
Owen, Richard, 16

Paley, William, 4, 8, 9, 15, 16
Parker, G. E., 206
Parker, Larry, 221
Pasteur, Louis, xviii, 87
Pastorius, Francis D., 175
Patterson, John, ix
Paul, Saint, writings of, 152
Pfeffer, Leo, 171
Philosophy of knowledge, xii–xvii, xxii,
  14–17, 37–43, 54–56, 83, 85–86, 94,
  100, 101, 124, 127, 159, 213–15
Popper, Karl, xvi, 41
Prigogine, Ilya, 24
Probability, xviii, 95–96, 101, 130–33, 215
Process theology, 156–59
Pro-Family Forum, 21
Progressive creationism, 154–55
Ptolemy, Claudius, 85
Public education, viii, xx–xxii, 19–21, 23–25,
  27, 34, 35, 55, 150, 164, 170–222
Pulaski County (Arkansas) Special School
  District, 216
Punctuated equilibria, 23, 77, 111, 213
Puritanism, 163, 172, 173, 176

Quakers, 173, 189

Radiometric dating, xiv, 27, 58, 63, 74,
  81–82, 126, 216
Ramm, Bernard, 154–55
Rauschenbusch, Walter, 165
Ray, Robert D., 189, 914, 196
Reader's Digest, 218
Reagan, Ronald, 37
Rehnquist, William H., 185, 186
Robinson, John, 163
Royal Society of London, 16, 17
Ruby, Jack, 38

Satan. See Devil
Sauer, Christopher, 175

Schlatter, Michael, 173
Schwarzschild, M., 54
Scopes, John, 3, 19, 170, 181
Seagraves, Kelly L., 23, 207
Seventh-Day Adventists, 183, 184
Sloan, Robert, 104
Slusher, Harold S., 24, 206
Smith, William, 74
Smithsonian Institution, 21
Society of Biblical Literature, 139
Speiser, E. A., 139
Spencer, Herbert, 7, 9
Stace, W. T., 32, 33
Stellar evolution, xvii, 25, 46–54, 127
Steno, Nicholas, 57
Sternberg, Saul, 38, 40, 41
Stewart, Potter, 184, 185
Strong, Augustus, 153, 154
Students for Origins Research, vii, viii, 190

Taylor, Raymond, 197
Teilhard de Chardin, Pierre, xx, 157–58
Tennessee Supreme Court, 181
Theistic evolution, 35, 155–56
Thermodynamics, 10, 17, 30, 111, 127,
    129–31, 154
  age of earth, 11, 12
  as defense of Bible, vii
  contrasted with kinetics, 133, 134
  evolution of life, xix, 11, 12, 31, 42
  origin of life, xviii, xix, 24, 90, 91, 96,
    101, 132–34
  origin of universe, 127–29, 134

Thomas, Curtis, 209
Time, 199
Tinkle, William, 21

Union of American Hebrew Congregations,
    187
United States Constitution, 170, 171,
    177–79, 181–87, 191–93, 210, 219–22
United States Supreme Court, 20, 170, 171,
    177, 181–87
Universe
  age, 46, 47
  origin, 14, 127–29, 134, 211, 212
University of Arkansas at Little Rock, 217
University of Central Arkansas, 217
Usher, James, 30

Wade, M., 80
Wallace, Alfred Russel, 7, 9, 87, 107
Ward, Nathaniel, 172
Watson, James, 89
Watson-Crick rules, 96, 97
Whitcomb, J. C., Jr., 60, 67, 70–72
White, Byron R., 184
Whitehead, Alfred North, 156
Wickramasinghe, Dr. (witness), 216
Wilberforce, Samuel, 16
Wilson, Marianne, 216–19
Wysong, W. L., 206

Young, W. A., 209

CONTRIBUTORS

DAVID B. WILSON is associate professor of history and mechanical engineering at Iowa State University. He received the B.A. degree in physics from Wabash College and the Ph.D. degree in the history of science from The Johns Hopkins University. He has taught at the University of Oklahoma and served as archivist for scientific manuscripts in the Cambridge University Library. His research and publications deal with the history of physics and connections between science and religion in the Victorian period.

WARREN D. DOLPHIN received the Ph.D. degree from Ohio State University in 1968. He taught and did research at the University of Maine at Orono for two years before moving to Iowa State University where he is professor of zoology and executive officer of biology. His research interests are in protozoan physiology and educational psychology. He has published papers in both areas as well as laboratory manuals for use in college biology courses.

DONALD E. BOLES is professor of political science at Iowa State University. He received the B.S., M.S., and Ph.D. degrees from the University of Wisconsin. He has published a book on religion and the public schools.

JOHN BOWER is associate professor of anthropology at Iowa State University. He received the B.S. degree in geology from Harvard University and the M.A. and Ph.D. degrees in anthropology from Northwestern University. Specializing in prehistoric archaeology, he has had field experience in excavations and artifact studies in Africa, Arizona, Illinois, and Iowa and teaches a graduate-level course on primate evolution. His doctoral dissertation was based on study of earliest stone tools from Olduvai Gorge, Tanzania. He is the recipient of numerous fellowships and grants and author of about thirty books, articles, and reviews.

HENRY A. CAMPBELL received the A.B. degree from the University of Nebraska-Omaha and the M.Div. degree from Federated Theological Faculty of the University of Chicago. He served the First Congregational Church in Moorhead, Minnesota, for eleven years and the United Church of Christ-Congregational in Ames, Iowa, for eighteen, before being called in 1985 to Parkview United Church of Christ in White Bear Lake, Minnesota.

ROBERT H. CHAPMAN, before his death in 1984, was assistant professor of botany at Iowa State University.

HUGO F. FRANZEN is professor of chemistry at Iowa State University. He received the Ph.D. degree in physical chemistry from the University of Kansas. His thesis research was in the area of application of thermodynamics to high-temperature vaporization reactions, an area in which his research group at Iowa State University has been active for eighteen years. He has taught thermodynamics at the graduate and undergraduate level and has published articles on the interpretation and teaching of chemical applications of thermodynamics. He has coauthored an introductory textbook on thermodynamics.

JACK A. GERLOVICH is the state science consultant for the Iowa Department of Public Instruction. He is president of the National Council of State Science Supervisors, and the Iowa Council of Science Supervisors, a Fellow of the American Association for the Advancement of Science, the Iowa Academy of Science, and serves on the boards of the National Academy of Applied Sciences and the Iowa Academy of Science. He is a former high school and college science teacher, author and/or coauthor of five science safety books and fifty professional science/education articles. He has served as a consultant to the American Chemical Society, the U.S. Office of Education, the National Institute of Education, Consumer Products Safety Commission, Federal Aviation Administration, and numerous private science/education companies. He is also a commercial airplane pilot.

BRIAN F. GLENISTER is A. K. Miller professor of geology at the University of Iowa. His degrees are in physics and mathematics from the University of Western Australia, in paleontology from the University of Melbourne, and in geology from Iowa. He has taught in Australian and American universities and served as chairman in the Department of Geology at Iowa. The author or coauthor of more than fifty books and scientific articles, he does research in invertebrate paleontology and biostratigraphy. He serves as a geological consultant in the petroleum industry, and is current chairman of the International Union of Geological Sciences Subcommission on Permian Stratigraphy.

PAUL W. HOLLENBACH is chair of the Religious Studies Program at Iowa State University. He received the Ph.D. degree in biblical studies from Drew University. His research has focused primarily on the life of Jesus, and he has published on this subject.

A. DAVID KLINE is associate professor of philosophy at Iowa State University. He studied biology at Wake Forest University from which he received the B.S. degree, and after serving in Vietnam as a medical service officer, he studied philosophy of science at the University of Wisconsin, from which he received the Ph.D. degree. His current research concerns time, causation, and use of the mathematical theory of communication to model several philosophically interesting concepts. He has edited two books and published over a dozen articles.

JANE L. RUSSELL received the Ph.D. degree from the University of Pittsburgh and did postdoctoral work at the University of Pittsburgh's Allegheny Observatory. She taught astronomy at Iowa State University before assuming her current position as assistant astronomer at the Space Telescope Science Institute in Baltimore, Maryland.

BRENT PHILIP WATERS is chaplain to the University and Director of the J. W. and Ida M. Jameson Center for the Study of Religion and Ethics at the University of Redlands, Redlands, California. He received the B.A. degree from the University of Redlands and the M.Div. and D.Min. from the School of Theology at Claremont. An ordained American Baptist and Disciples of Christ minister, he has published articles on religion, science, and technology in a variety of religious periodicals.

STANLEY L. WEINBERG received the B.S. degree from City College of New York and the M.S. degree from Northeast Missouri State. A biology and science teacher and supervisor for most of his career, he has done research in animal behavior at the American Museum of Natural History, in virology and tissue culture at Yale Medical School and Sloan-Kettering Institute, and in food technology at John Morrell & Co. Research Laboratory. His publications include several high school and college biology textbooks, and numerous journal, magazine, and newspaper articles. He has held office in the National Association of Biology Teachers and the New York Biology Teachers Association. He serves as Coordinator, Iowa Academy of Science Panel on Controversial Issues, and National Coordinator, Committees of Correspondence (in defense of evolution).

JOHN H. WILSON received the Ph.D. degree in biochemistry and genetics from the California Institute of Technology. After postdoctoral studies at Stanford University, he went to Baylor College of Medicine, where he is associate professor in the Department of Biochemistry. He has contributed actively to the scientific literature in the areas of molecular genetics and, in particular, the mechanisms of genetic recombination in higher organisms. In addition he has coauthored introductory textbooks on biochemistry, molecular biology, and immunology.

BRIAN J. WITZKE is a research geologist at the Iowa Geological Survey and adjunct assistant professor of geology at the University of Iowa, from which he received the Ph.D. degree. Before coming to Iowa as a teaching-research fellow, he was graduated with honors in geology from the University of Wisconsin-Milwaukee. He has authored papers in paleontology, stratigraphy, and paleogeography and has additional research interests in sedimentary petrology and biostratigraphy.